Building in Visual Concrete

*Illustrated by reference to
100 notable structures in the
British Isles and Overseas*

Adapted from the original German text of

Professor Erwin Heinle
*Professor of Architecture at the Stuttgart
Akademia and one of the best-known architects
in West Germany*

and

Professor Max Bächer
*Professor of Design and Space Treatment at
the Technical University of Darmstadt*

by the Editorial Staff of the Publishers
with the linguistic assistance of
Mr. Joseph Berger, F.R.I.B.A.

and

Mr. S. V. Whitley, B.A.

and under the technical guidance of
Mr. K. H. Brittain, F.I.C.E., F.I.Struct.E.
(Group Technical Consultant)

and

Mr. A. L. Edge
of the Blue Circle Group, London

Technical Press, London

SBN:— 291 39299 7

Set in Monotype Times Roman and printed in England by
The Whitefriars Press Ltd., London and Tonbridge

Bound by G. & J. Kitcat Ltd., Shand Street, London, S.E.1

Preface

THE TERM *visual concrete* is used in this book to describe any concrete surface which has been deliberately planned to remain visible to the beholder after the building of which it forms part has been completed. The term is intended to embrace both concrete cast on the site itself and prefabricated concrete facing components cast elsewhere. It covers 'exposed concrete' whose surface has been left untouched after dismantling of the formwork, and concrete whose surface has been subjected to raking, hammering, sandblasting or tooling of any other kind. It covers also concrete whose surface has been removed altogether (by washing, brushing or etching with acid) so as to leave exposed the carefully-selected aggregate used in the making of the concrete itself.

No term with so extensive a technical meaning appears to be currently in use on either side of the Atlantic. The nearest approximations are probably *architectural concrete*, *visible concrete* and (perhaps especially) *exposed concrete*; but to all of these objection is taken by many architects and other users of concrete as a building material. *Architectural concrete*, for instance, besides being ponderous, is disliked because it fails to indicate that the material is there to be looked at. *Visible concrete* seems too much like a statement of the obvious. *Exposed concrete* contains for some an unwanted suggestion of deliberate exposure to the elements, while to others it seems an unsatisfactory way of describing a surface which may often be not so much 'exposed' as removed altogether to display something lying beneath it.

So when an accurate English translation of the German term *Sichtbeton* (with its accepted French equivalent *béton apparent*) was needed, a new expression had to be sought. Several alternatives were considered—among them *display concrete*, *revealed concrete* and even *seen concrete*—but all met with objection of one kind or another; and eventually *visual concrete* was chosen as being, on balance, the most generally acceptable. Though it involves using the word 'visual' in a rather unusual sense, the term is brief, descriptive and suggestive of no motive foreign to the intentions of the architect. It is therefore put forward in the hope that it may be found a useful generic term having the wide meaning defined.

BUILDING IN VISUAL CONCRETE is, in the main, an English-language version of a work entitled *Bauen in Sichtbeton*, published in 1966 by Julius Hoffmann Verlag of Stuttgart. This book combined a pictorial and technical account of some of the outstanding work done in visual concrete all over the world with an essay by Prof. Max Bächer exploring the intellectual and aesthetic possibilities of the new material and a long chapter by Prof. Erwin Heinle discussing the many technological problems which arise in its use.

Both were highly qualified for their respective tasks, Professor Bächer being both a practising architect and holder of the Chair of Design and Space Treatment (*Lehrstuhl für Entwerfen und Raumgestaltung*) at the Technical University of Darmstadt; while Professor Heinle is not only Professor of Architecture at the Stuttgart *Akademia* (the approximate equivalent of London's Royal College of Art), but one of the best-known architects in the Federal Republic. Among the notable buildings for which he has been responsible are several hospitals and large industrial plants in South Germany, and the Television Tower and Parliament Buildings for the *Land* of Baden-Württemberg, both in Stuttgart. He is currently at work on the Olympic City which is being built at Munich for the Games of 1972.

In translation, Professor Bächer's essay has (with permission) been freely rendered and much condensed, while to Professor Heinle's chapter have been added both detail and comment reflecting current British constructional practice. To both gentlemen, and to their German publisher Herr Kurt Hoffmann, thanks are due for their courtesy in allowing such changes to be made and for their help in resolving difficult technical points arising across the language barrier.

The initial task of translating the German text into English was undertaken with the linguistic assistance of Mr. Joseph Berger, a Fellow of the Royal Institute of British Architects whose mother-tongue is German, and of Mr. S. V. Whitley, BA, an official translator in the Navy Department of the Ministry of Defence. The text was then submitted (by permission of Sir John Reiss, Chairman) to the Technical Department of the Blue Circle Group, where the Group Technical Consultant, Mr. K. H. Brittain, FICE, FI.struct.E, and especially his Assistant, Mr. A. L. Edge, spent many hours in checking the technical soundness of the translated text and adding to it many parallel details of British practice.

Chapter 8 of the book—a new chapter in which an account is given of some outstanding work in visual concrete in the British Isles—though compiled in the offices of THE TECHNICAL PRESS, owes much to the courtesy of the Cement and Concrete Association and of Mr. George Perkin, ARIBA, editor of the Association's admirable journal, *Concrete Quarterly*. Mr. Perkin not only arranged for the loan of the great majority of the photographs used from the Association's Research Station at Wexham Springs, near Slough, but also allowed great use to be made of the text which had accompanied these photographs in the original issues of *Concrete Quarterly*. Miss Peggy Pearce (photo librarian at Wexham Springs) spent much valuable time looking out and despatching the photographs requested.

The publishers are grateful also to the many clients, architects and photographers who have allowed them use of the photographs reproduced throughout the book—and especially to Sir Hubert Bennett, FRIBA, FSIA, Architect to the Greater London Council, for the loan of several fine photographs of London's new South Bank Arts Centre taken by the Photographic Unit of the GLC's Department of Architecture and Civic Design.

It is not easy to remember today that the use of concrete as an architectural material in its own right is a development entirely of the twentieth century. Before 1906, the date of Frank Lloyd Wright's pioneering "Temple" in Oak Park, Illinois (*illustrated on page 22 of this book*), concrete was used only as a structural prop for more traditional building materials. When (rarely) it was used as the sole constructional material in the erection of a building, it was unhesitatingly deemed unfit to be left exposed to the eye of the beholder and was hidden under varying thicknesses of stucco, plaster or paint.

Now all that is changed. Since the early 1900s the architectural use of concrete has made immense strides, and great structures in this most flexible of all building materials are going up all over the world. Arguably, indeed, the development of visual concrete is the most important contribution which this century has made to the history of architectural thought—though, like all radical innovations, it has never ceased to provoke vigorous controversy on grounds of aesthetic taste. It is the aim of this book to illustrate the exciting challenge which the new material makes to the creative imagination of the architect, and to describe some of the solutions which have been found to the problems it poses to the technical skill of the structural engineer and contractor.

Table of Contents

Chapter 1

The design and construction of buildings in visual concrete

Adapted from an Essay *Hinweise für Planung und Ausführung von Sichtbetonbauten*

by Professor Erwin Heinle

The recommendations put forward in this Chapter are written by an architect for architects. They claim neither to be scientifically correct in every particular, nor to comply exactly with the fully-accurate definitions given in established Standards. They are intended, rather, to serve as an introduction, and to provide practical answers to the types of problem which arise in practical work.

The nature of visual concrete

By "visual concrete" is meant a concrete whose surface is deliberately planned to remain visible after the building is completed. The original concrete structure must therefore be neither painted nor rendered. In the wider sense, washed concrete and concrete surfaces treated as if they were stone (e.g. bush-hammered) may also be counted as visual concrete.

Buildings in visual concrete require a great deal of pre-planning. Every factor affecting form and surface structure must be thought out in advance, as must the provisions for later installations and similar additions. Changes in the finished building are often from the practical point of view impossible, for they cause a great deal of work and always remain apparent.

Definitions

By *dense concrete* is meant the usual form of concrete having a specific weight of 1900 to 2500 kg/m³. (Freshly-made concrete of this grade normally attains a specific weight of 2300 to 2400 kg/m³.) *Lightweight concrete* has a density of 1800 kg/m³ and below. *Heavy concrete* has a specific weight of 3000 to 5000 kg/m³, and results from using heavier aggregates like barytes or scrap-metal.

The qualities of concrete are defined by reference to its compressive strength. In Germany, concrete of quality B225 must be able to resist a pressure of at least 225 kp/cm² 28 days after being poured. The equivalent British "standard mix" (CP 114—Grade 21) has a compressive strength of 21 N/mm² after 28 days.

For building above ground the following concretes are suitable: Standard mixes, Grades 21, 25·5 and 30. For normally stressed components Grade 21 is normally used; if the stressing is higher, Grades 25·5 or 30 are required. For good visual concrete, Grade 30 or an even higher grade of concrete should always be used.

The composition of concrete

Concrete is an artificial form of stone consisting of gravel or crushed rock and sand, bonded together by cement. It possesses (according to quality) high compressive strength, but practically no tensile strength. For concrete to acquire this latter quality, it is necessary to introduce steel reinforcement. The result is *reinforced concrete.*

Stressed concrete is produced by applying special tension to the steel reinforcement after the hardening of the concrete. This pre-stressing sets up forces in the concrete which act to oppose stresses applied to it later on. This has the effect of raising considerably the load-carrying capacity of stressed concrete.

Concrete placed into the formwork on the site is called *in-situ concrete.*

Prefabricated components are made either in the concrete works or in special installations on the building site itself, and are then stored in their hardened state until required for assembly. Prefabrication of this type guarantees higher and (even more important) uniform quality. It also enables work to continue through the winter.

The characteristics of finished concrete are governed by the following factors:
(a) The type and quality of the cement used.
(b) The type and grading of the aggregate used.
(c) The proportions of cement and aggregates in the mix.
(d) The water/cement ratio of the mix.
(e) The consistency of the concrete.
(f) Its workability and density.
(g) The ambient conditions during placing and hardening.
(h) After-treatment.

Cement

Cement is a hydraulic binder which hardens in air or water. The various types of cement, and the qualities which each must possess, are specified in National Standards in every leading industrial nation in the world. (The relevant British and U.S. Standards include: BS 12, BS 4027 and BS 146; ASTM C 150, Types 1, 2, 3 and 5).

Cement absorbs moisture from the air. The finer the particles, the more absorbent the cement. This affects the period for which cement can be stored without deterioration. The finest grades should not be stored for longer than a month; but this period can safely be doubled with cements of coarser particle size.

When cement sets, a form of crystal results from the hydration of the cement powder. It is this crystal which is the real source of the hardness of concrete, and also of the largest variations in its volume. In the literature on the subject, the composition of the cement and its *granular fineness* are quoted as the principal factors in determining how it will set. The higher the quality of the cement, the quicker will concrete made with it harden. The coarser the cement, the lower will be the early strength of the concrete (given otherwise uniform conditions).

The *colours* of the various types of cement are greys of sharply distinctive tone. Some works supply cement having a pronouncedly greenish hue; for others a blue-grey tint is characteristic, and so on. There is also white Portland cement, which is about twice as expensive as grey Portland cement; but this has no great effect on the price of concrete produced by using it.

Aggregates

The aggregates used must be hard, and must contain no clay, shale or decaying stone. Aggregate of primary rock is generally more suitable than limestone, because it absorbs less water and is more resistant to frost. Crushed stone is acceptable, provided the stone itself is hard; but it can absorb more water, so the quality of concrete made with such aggregate is more difficult to control. Organic impurities such as humus, coal particles and certain sulphur compounds are harmful to the making of good concrete.

To give concrete its necessary density (strength, appearance and durability) the *grading* of the aggregate used must be correct. For building above ground with reinforced concrete, the maximum size of aggregate is usually up to 30 mm (1·25 in). The grading is expressed in sieve ranges laid down by British Standard BS 882 (ASTM C33 is the comparable U.S. authority). For high-quality concrete, the aggregates should be supplied in separate sizes, and should be batched separately. During concreting, the site supervisor should repeatedly check that the grading of the aggregate being used remains as specified.

To achieve dense concrete, the content of very fine sand or stone powder under 0·2 mm in diameter should be relatively high—about 380 to 420 kg per cubic metre of concrete (640–700 lb per cubic yard). To achieve this, either extra-fine particles (stone powder or very fine sand) are added, or more cement is added to the mix. Like the cement itself, these particles of stone powder can have a considerable effect on the colour of the concrete. They also to some extent bring about changes in its volume. (*Note by British Editor—Specialist advice on this point should be taken.*)

Mixing water

The water used must always be potable (i.e. free from oil, fat or sugar). When not taken from the public supply, it should be tested.

The ratio by weight of water to cement used is called the *water/cement ratio.* For good visual concrete, this factor should be about 0·4–0·5. The consistency of the freshly-made mix depends in no small measure on the water/cement ratio.

With a high water content, the tendency of concrete to "bleed", shrink, crack or creep increases; strength and resistance to frost diminish; and both heat transmission and condensation may be affected. For the chemical process of hydration to be completed, a water/cement ratio no higher than 0·25–0·28 is necessary. Extra water makes the concrete more workable; it will evaporate during the curing and drying-out processes.

An excess of water affects the characteristics of the concrete but makes it easier to mould. The choice for the man in charge is a difficult one. It is generally a sound rule to adhere strictly to the specified water/cement ratio. If the matter is left to instinct, water content always tends to be too high.

Variations in water content affect the porosity of the concrete, and so give rise to variations in its colour. The higher the humidity of the air, the more these variations will become visible.

Admixtures

Opinions about admixtures differ considerably, the principal reasons being that preparations vary in their effects and that their composition is frequently unknown. They are added in small percentages to the mix. The following main types emerge:

Retarders and *accelerators,* both of which affect the speed of the chemical process;

Water-reducing agents and *plasticizers,* which reduce the water content required and make the fresh concrete more workable;

Air-entraining agents; and

Waterproofers.

Water-reducing agents and plasticizers lend themselves especially to the manufacture of visual concrete. They allow the water content to be reduced, and lessen the risk of the concrete segregating and "bleeding". A concrete technologist should be consulted to advise on admixtures.

The colours of hardened concrete

These depend, first, on the colour of the cement itself and, second, on the colour of the fine aggregates used. These latter range from pure white to dark grey, with greenish, bluish and yellowish tints in between. The smoother the surface, the lighter in colour the concrete will appear—an effect which remains after waterproofing. Rough surfaces appear darker because of the shadows cast by their texture; they also get dirty more quickly.

The desired colour should be established on a *sample panel*. In order to avoid particularly bad colour variations, the same source of supply of cement should be used throughout a given job.

White concrete is made with white Portland cement. If the shuttering is made smooth and compact and if a water-repellent is applied after striking, the concrete will emerge, and will remain, nearly white.

Colouring is also possible by adding pigments to the concrete. Such pigments need to be fast and alkali-resistant. Titanium dioxide and zinc sulphide make the concrete lighter in colour; iron oxide makes it darker (yellow, red, brown or black). Chrome oxide (giving a green hue) and manganese (giving a blue tone) can also be used. Used as a colloid, they give a strong colouring effect. The quantities added must be related to the weight of the cement. Between 3% and 5% by weight is sufficient. Coloured concrete is not often used for *in situ* work, for it is expensive and difficult to make uniform; but it is suitable for pre-fabricated elements.

Special types of concrete

Prefabricated units require an especially dense consistency, which is achieved by compacting the concrete on vibrating tables. (*Note by British Editor—ALL visual concrete requires vibration.*)

Even with normal cements, it is possible to achieve compressive strengths up to some 90 N/mm²; but such very high strengths are merely laboratory results and cannot yet be used economically in practical building. Indeed, German building regulations do not yet allow concrete with such values to be used at all.

Still higher compressive strengths can be achieved if measured quantities of polyester resin are added to the cement or if it is treated with magnesite—strengths as high as 120–200 N/mm²; but they, too, have not yet proved economically possible on the building site.

Steam-curing is sometimes used in prefabrication. The super-heated steam speeds up the hardening process and increases the early strength of the product. It is possible to achieve compressive strengths of up to 80 N/mm² (about five tons per square inch) after 12 hours of curing.

Waterproof concrete requires a capillary structure different from that of ordinary concrete. If the water/cement ratio is kept down and suitable plasticizers are added, it is possible to achieve a practically non-porous concrete. The author has not, however, heard of any example of completely waterproof concrete used over large surface areas.

For concreting in soil which contains sulphates, either solids or in solution, *sulphate-resisting Portland cements* must be used containing no tricalcium aluminate. Similarly, concrete which is to be exposed to aggressive fumes must be specified as such. In all such cases, cement chemists should be consulted.

With expanded clay and slate aggregate, *lightweight concretes* have been produced which, having densities of only 1300–1600 kg/m³ (80–100 lb/cu. ft), can show compressive strength of up to 45 N/mm². Such lightweight concrete has been used in many buildings where the surface has been left exposed.

To achieve particular effects, e.g. with exposed aggregates or by colouring, a layer of special *facing concrete* can be placed in front of the structural concrete. This is an intricate job if it be attempted when the formwork is vertical; but it is simpler (and therefore more often used) for horizontally-cast prefabricated units.

Whereas in normal dense concrete the grading of the aggregate is continuous, *no-fines concrete* is made with aggregate consisting of single-sized particles all graded in the 15–30 mm ($\frac{5}{8}$″–$1\frac{1}{4}$″) range. The aim is not to achieve an optimum density of aggregate, but a loose structure in which the gravel or broken aggregates are bedded in cement paste but the interstices between them are left unfilled.

Ready-mixed concrete is not a particular kind of concrete. The expression merely means that the concrete is not made on the site, but in a factory from which it is delivered fresh to the site. Its advantages are a very even mix; storage space is saved; and there is no need to find room for bulky mixing plant on site. Its disadvantages are that special unloading equipment is needed on site, with resultant need for organizing deliveries. The transport itself can also present difficulties.

Concrete characteristics

For successful building with concrete, and especially with visual concrete, the special characteristics of the material must be borne in mind. Of its many qualities, perhaps the most important is its outstanding suitability for the construction of monolithic components of great size. Visual concrete, of course, in addition to supporting the structure of the building, must also measure up to demanding aesthetic standards.

Strength and plasticity

Obvious advantages of concrete in general (and therefore of visual concrete in particular) are its great strength, which allows the use of bold, slender components; the nearly unlimited range of forms which it makes possible, both in overall appearance and in detail; its variable surface texture; and its range of colours from white to strong, dark tones.

Durability

Other advantages of concrete are: It has good resistance to weather and frost (note, however, that there must be no honeycombing if frost damage is not to occur). It weathers well. With the passage of time it acquires an agreeable patina not unlike that of limestone. It offers good resistance to a polluted atmosphere if it is made very dense and with resistant types of cement.

If dust has a chance to settle on cornices, etc., there is always the danger that it will contain particles of sulphuric oxide which will discolour the elevation when washed down by rain; but the application of water repellents gives good protection against this danger for a number of years.

Concrete offers good protection against fire, provided that the proportion of reinforcement forms less than 2% of the cross-section. Its toughness and resistance to abrasion make for durability. Even if pieces of a concrete structure are knocked away, the effect on so homogeneous a material is less disturbing than it is on, e.g. stucco or paint.

Sound-proofing capacity

Being a heavy material normally used in considerable thicknesses, concrete is a good insulator against *airborne sound*. Its density, however, makes it an efficient conductor of *impact sound*. Good insulation against airborne sound is thus allied to appreciable sound conductivity, against which precautions need to be taken (*see below*).

3

Heat insulation and heat conservation

The *heat conductivity* of concrete is two to three times that of brick; it therefore often requires a measure of insulation against heat losses. Its *heat-storage capacity*, by reason of its heavy specific weight and usually large mass, is good. This enables concrete to act as an efficient heat exchanger between day and night temperatures.

Heat expansion

In concrete, the *coefficient of expansion* (which indicates the change in length caused by a difference of 1 °C in temperature) is approximately 10×10^{-6}, which is about the same as that of natural stone. The coefficient of expansion for aluminium is about $2\frac{1}{2}$ times that of natural stone, and the coefficient of expansion for plastics is 6 to 20 times larger.

In the case of stone or brick walls, the effects of expansion are taken up by the many joints and fissures between the structural elements, and normally remain minute and scarcely visible. But concrete is usually cast in large monolithic slabs. On the Continent of Europe, temperature differentials of 100 °C need to be allowed for (between the −30 °C of a cold winter's night and a surface heated by a midsummer sun to as high as 70 °C). With such a temperature differential, a 60 m (180 ft) long concrete wall will expand about 6 cm ($2\frac{1}{4}$"). To prevent deformation, either expansion joints or adequate pre-stressing are necessary.

Other changes in volume

In the process of setting, the volume of concrete begins to shrink while it is still in the plastic state. Cement-rich mixtures shrink most, and the rate of shrinkage increases with the rate of strength development. If the mass of concrete is not too great, the effects of shrinkage are normally not noticeable. It can, however, cause surface cracks.

Shrinkage is part of a complex chemical process in which the process of hydration releases heat. In thick walls, the temperature at the centre may reach 65 °C (149 °F). Sudden quenching of such a wall by hosing it with cold water could cause cracking, as could other methods of too-rapid cooling.

Finished concrete shrinks a little as it dries out and expands again as it becomes damp, especially when it is of a porous type. A high water/cement ratio and a high cement content increase these effects. The type of cement and aggregates used in its manufacture, and the conditions in which it is allowed to dry out, are also important.

Shrinkage cracks may develop if such a change of volume is impeded, e.g. by very rigid connection to the foundations. To avoid this, a water/cement ratio of less than 0·5 (preferably even lower) should be specified.

Concrete should always be covered with sheets of polythene when freshly struck from formwork.

Shrinkage cracks are basically defects of appearance, and can be optically overcome if they appear on strongly sculptured surfaces. But if they expand excessively, they can reduce both structural strength and impermeability, resulting in possible damage to the reinforcement.

In small building components, little or no such damage will occur if the methods of manufacture have been satisfactory.

Under permanent loading, concrete deforms and a defect known as *creep* develops. A high water/cement ratio and high cement content increase the tendency to creep. Too-rapid drying may also have the same effect. In slender vertical members creep is not noticeable, but horizontal members which have not been pre-stressed can bend.

Remedies are proper camber, correct reinforcement and pre-stressing; but such changes of dimension play a less important role when prefabricated elements are being used.

The term *bleeding*, or *water-rejection*, is used when some of the water incorporated in the mix oozes out on to the surface of freshly cast and compacted concrete. Such a process taking place inside the formwork may result in the uneven distribution of solid matter, and blemishes may appear in consequence on surfaces of smooth visual concrete. The tendency to "bleed" can be increased by mixes lacking in fine particles, by a too high water/cement ratio, or by the use of coarsely-ground cement.

Re-vibration of the concrete may enable the disturbing structural alterations to be corrected. Some admixtures also have the side-effect of reducing bleeding; but they need to be carefully chosen, for their effects differ.

Water-retention and saturation-point

The retention of *in-situ* concrete is relatively high, and a concrete structure retains water longer than does, for instance, brickwork. This gives rise to higher initial heating costs. It is not a factor of importance in the case of finished components, because they are generally dried out before installation.

The saturation point of concrete is only about one-quarter that of brickwork or lime rendering. This means that a concrete wall will absorb very little condensation. Kitchen and bathroom walls in concrete will therefore tend to be damp. But this characteristic results in good resistance to weather and frost, which can be increased by coating with water-repellent.

The design of buildings in visual concrete

Concrete buildings demand especially careful pre-planning. Not only the overall design, but also the *details* and the *service connections* must be thought out and finally decided before the first concrete is poured into the formwork. This applies especially to holes, chases and grooves for pipes, conduits and fixings. Those minor alterations during and after erection which are quite usual with traditional building materials are very wasteful in concrete, and in visual concrete very conspicuous also. They are frequently inadmissible, furthermore, for reasons of cost.

All complicated corners, points and connections should be worked out with the aid of *three-dimensional models* before casting begins.

For all larger contracts in visual concrete, the *contractor* chosen should have wide knowledge and experience of the methods involved. Smaller structures are easier to execute, but large surfaces in visual concrete require very careful formwork design and casting.

In the design, the characteristics of the various materials mentioned earlier in this Chapter need to be carefully borne in mind. In German law, at least, architect and engineer are held liable for any errors; and on this point the Courts in the Federal Republic are tending to take an ever-stricter view.

Specifications

More explicit and detailed specifications are called for in structures in visual concrete than in buildings constructed by traditional methods, in which much required knowledge and experience can safely be taken for granted. All concerned must clearly understand the full implications of working in visual concrete—not only employees of the principal contracting firm itself, but also of all sub-contractors. Special points to clarify are the precise chain of command, and the responsibilities of every party in respect of damage and defects.

In addition to the standard code of good building practice, the following items require specially close attention:

(a) The type and method of erection of the formwork—if possible with models.

(b) Detailed drawings of all formwork to be available.

(c) The type of cement to be agreed, and care taken that all cement used conforms strictly to sample in its characteristics.

(d) Sample panels and approval thereof to form part of the contract.

(e) None but officially approved types of admixture to be used.

(f) Ties and spacers (whether to be exactly as specified, or whether to be calculated on the basis of the original quotation).

(g) Methods of compaction (with special emphasis on the longer compaction period required by visual concrete).

(h) Dismantling of the formwork—how soon after placing is it to be begun?

(j) Treatment of the finished concrete.

(k) Protection of the finished concrete, from first exposure to handing-over time, with extra allowance for continued protection of all visual-concrete surfaces, arrises, corners, etc. after handing-over.

It is recommended that a site foreman with some knowledge of the special problems of visual concrete should always be nominated.

Heat insulation

Where (as in Germany) outside temperatures in winter can fall well below $-15\,°C$, the degree of heat insulation provided by a concrete wall is insufficient for inhabited rooms. In Climate Zone II ($-15\,°C$) such a wall would need to be 95 cm thick, and in Climate Zone III ($-18\,°C$) 115 cm thick. Walls of such thickness would require considerable energy to heat them. It is therefore usual to keep wall thicknesses to the statistically correct measurements and to achieve the necessary heat insulation by means of additional layers of material such as woodwool/cement panels (*Heraklith*), foamed plastics, glass and mineral fibre, foamed glass or cork.

When the insulation is fitted on the *inside* of the wall, as is usual, it is essential to protect it by means of a waterproof membrane against penetration by humidity, which would reduce the effectiveness of the insulation and could lead to structural damage. A covering of rendering, lath-work or panelling can be used for this purpose, but care must be taken not to damage it in any way. Even small holes in a waterproof membrane can lead to trouble.

If applied to the *outside* of the wall, the heat insulation must be protected by a weather-proof and damp-proof coating.

Should it be desired to preserve the characteristic appearance of visual concrete on *both* sides of the wall, the insulating material must be sandwiched into the wall itself. For this only completely stable materials, impervious to damp and to atmospheric humidity and sufficiently rigid to withstand damage by vibration, can be used.

Sandwich walls of this type are very difficult to cast in one lift, for there is inevitable trouble with the formwork and much risk of damage to the insulation material. Usually, therefore, one skin is cast, the insulating material is fixed, and the second skin is cast on to it. But this method is not cheap.

It is important to remember that any connections between the two concrete skins will act as heat conductors and will considerably reduce the degree of insulation provided. If only 15% of the surface area of the walls is interconnected by such cold-bridges, the total insulating effect can be reduced by as much as 50%.

In such buildings as churches or assembly rooms which are used only for short periods, it may be more economical to omit the heat insulation altogether and to rely instead on larger heat-producing surfaces or more warm-air curtains inside.

One psychological effect to be considered is that most people regard concrete as a "cold" material. Care should be taken to counteract this effect by means of especially effective insulation, by generous heating arrangements, and by the use of contrasting surfaces of warm colour in timber, textiles, flooring, etc.

Condensation

This is a problem closely linked with that of heat insulation. When the air inside a room is cooled to a degree where its humidity reaches saturation point and can no longer be kept in suspension, condensation will occur on the cold walls.

Porous materials used in the decorative treatment of walls and ceiling can help to solve the problem, for they are capable of absorbing considerable quantities of water which they later give back to a warmer atmosphere. The moisture also moves as dampness through the wall to the colder exterior.

Compared with brick, however, concrete can absorb only small quantities of water and is a poor conductor of vapour. It is not damaged by condensation, as is plaster.

Black mildew can, of course, always appear on walls exposed to condensation, and water running down them can damage timber members and floors. In solidly-built industrial buildings, condensation need not restrict the use of the rooms in which it occurs; but in all cases adequate provision must be made in the design for collecting and channelling off the resulting water.

In the climate of Germany, provision must always be made against condensation in permanently inhabited rooms built in concrete; but the appearance of condensation is difficult to to predict from mere data about the wall section itself. Conditions on site are usually so complex that predictions based on laboratory work fail to apply—in the good sense as well as in the bad. There is no substitute for sound practical experience, and the constructional engineer must be able to take into account in his calculations the whole environment of the building and the conditions ruling during the course of its erection.

In the United Kingdom, a factor to which great weight is attached is the necessity of providing adequate ventilation in situations where condensation is likely to take place (*see Building Research Station Digest 91, 2nd Series*).

Sound insulation

Concrete walls, floors and ceilings having a thickness of 24 cm ($9\frac{1}{2}''$) give protection against *airborne sound* of up to 5 decibels. This is not enough, however, for residential party walls because of the longitudinal sound-conductivity of the concrete.

The high conductivity of *impact sound* by thin concrete members is best counteracted by soft floor coverings or floating screeds, so that the sound of footsteps or louder impact noises is not transmitted to the concrete itself. It is also, of course, essential that plumbing noises be kept to a minimum, and that such matters as "silent" light switches and curtain runners be attended to.

Elastic separation joints which break the path of the sound conduction can often be effective in preventing the spread of impact sound through a whole house; but in that case all *sound bridges* such as nails or rubble connecting adjacent concrete members must be carefully avoided.

Cracking of concrete

In addition to the thermal stresses which, as has already been mentioned, may affect concrete structures, cracks in the concrete can also be caused by unequal settlement along the length of the structure. To meet this danger, *expansion joints* are necessary; but they should only be inserted after close consultation with the construction engineer. Uncontrolled expansion joints can impair the strength of the construction, damage the reinforcement, give rise to saline efflorescence and expose the structure to damage by frost.

If expansion joints are considered undesirable for either structural or aesthetic reasons, *pre-stressing* can be used, in cooperation with the construction engineer. Normal (unstressed) reinforcement to counteract shrinkage has not, however, proved successful. (*Note by British Editor— Experience in the U.K. suggests that, given correct distribution of unstressed reinforcement, shrinkage cracks do not become large enough to be visible.*)

In normal circumstances, all reinforced external walls of *in-situ* concrete—balcony slabs, balustrades, canopies, etc.— should have expansion joints every 3 to 6 metres (10 ft to 20 ft) along their length, instead of the 10 to 15 metre (33 ft to 50 ft) spacing considered appropriate with traditional materials. It is also sometimes necessary to provide horizontal expansion joints, e.g. between foundations and rising walls.

Concrete roofs should not be exposed to the direct heat of the sun, but should be insulated even if the use to be made of the space below (e.g. for a garage) would not normally justify such a precaution. Expansion joints are essential here also— again after consultation with the construction engineer.

Cracks arising from the technological nature of concrete (such as the shrinkage mentioned earlier) must also be dealt with. Provided these cracks are not deep, they will only affect the appearance of the structure, whose rough, vigorous surface can often disguise them as light and shade. Larger and deeper cracks, however, can give rise to the kinds of damage already mentioned. Such cracks can be prevented from forming by using a cement having a low shrinkage value; by using a not-too-rich mix for the concrete and a low water/cement ratio; by careful placing and curing of the concrete; by skilful striking of the formwork; and by correct after-treatment (*see below*).

Large unbroken areas of concrete tend to lead to shrinkage cracking, as does the tension set up in those parts of the structure where the cross-sectional area is locally reduced (e.g. by apertures in the structure itself). Insulating materials inserted between two skins may cause heat to be stored in the outer skin, and so give rise to differential expansion and resulting cracks.

Cracks can often be localized (i.e. prevented from spreading) by the reduction at planned intervals of the cross-sectional area of the concrete.

Haircracks disfigure a surface by collecting and retaining dust and dirt. They can be removed by sand-blasting, but this changes the appearance of the surface both in colour and texture.

All larger cracks should be carefully filled with a mastic filler so as to prevent further widening of the cracks by water and frost and consequent additional damage. But if the worst comes to the worst, there is no alternative to cutting out the affected parts.

Cracks will not normally be a hazard in building with prefabricated concrete components. The individual components are much smaller, and their manufacture can be kept under strict control. Storage conditions during controlled maturing are also far more favourable than they can ever be on a site.

The formwork and its planning

The design of the formwork is normally the responsibility of the contractor or of the structural engineer; but because where visual concrete is concerned the formwork has a decisive influence on the character of the whole building, it behoves the architect himself to keep a careful eye on such details as board widths and joints, and the exact way in which the boards are erected.

The cover to the steel reinforcement needs to be greater in buildings in visual concrete than it does in ordinary concrete structures. It varies with the maximum size of the aggregate. The cover should always be 5 mm thicker than the maximum aggregate size so that individual pieces of aggregate cannot get stuck between formwork and reinforcement and so cause uneven compaction. Thus with the normal maximum aggregate size of 30 mm, cover should be 35 mm. Surfaces which are later to be worked by tooling, grinding or polishing (*see below*) need thicker cover still.

It may even pay sometimes to work with smaller-sized aggregate. This means that the cover can be reduced, with a resulting saving in weight and less heavy reinforcement; but the penalty will usually be a reduction in loading strength. Maximum aggregate size, however, should never be reduced below 20 mm in structures exposed to the open air.

Formwork materials

The visual-concrete surface to be produced is the *negative copy* of the formwork itself. Suitable materials for the formwork are thus:

Rough unplaned boards (sand-blasted, if desired);
Planed boards;
Panels of plywood or of other smooth-surfaced materials (having, if so desired, a plastic finish);
Sheet-metal;
Asbestos-cement sheets.

The boards used may be fir, Scots pine, Carolina pine or pitch pine. They should be well-seasoned and relatively free from knots, above all loose ones. Their capacity to absorb mould oil depends on the density of the wood and its resin content. Variation in these two characteristics can cause spots. New boards should be oiled at least twice.

How many times formwork should be re-used, and whether one does better to use fresh boards or fresh boards "pre-aged" by treatment with cement grout, is a matter of dispute. "Pre-ageing" is, in the author's experience, of problematical value, because when the boards dry cement particles tend to crumble off them.

The finer and lighter the visual-concrete surface is intended to be, the smoother and thicker is the formwork required. A rough surface tends to conceal surface blemishes by its darker tones. For entirely smooth surfaces, accurately planed boards are generally to be recommended. Conversely, sand-blasted boards impart to the concrete a vigorous image of the texture of the boards themselves.

United Kingdom practice tends to avoid concrete surfaces which can be described as "very smooth off the form". Such surfaces have too often given rise to a high incidence of blemishes and of such defects as hair crazing.

Formwork faced with hardboard or plywood requires solid support. If hardboard is to be re-used several times, its thickness should be at least 4 mm. Otherwise 2·5 mm is normally enough. Sheets of oil-impregnated hardboard are easier to shape. Plywood panels 4 to 5 mm thick are more expensive than hardboard, but they last longer. If desired, they can be coated with a plastic lacquer.

Care must be taken to minimize swelling or shrinking of the

formwork. It should be well damped before the concrete is poured, for this makes striking much easier.

Strongly profiled wall surfaces can be achieved by using formwork containing, e.g. triangular battens or chamfered edges. All profiles must be designed for easy striking so as to minimize damage. In all such formwork the essential concrete cover of the reinforcement must be maintained, the most deeply indented spots being the critical ones. For this reason, powerful profiling can involve considerable increases in weight. For this reason also, prefabricated components with strong profiles are generally cast hollow.

Complex shapes such as sculptured models can be made of plaster or of a stable foamed plastic embedded in the formwork. For sculptural work, moulds reinforced with glass fibre are sometimes used.

Sharp right-angled arrises are liable to break off when the formwork is struck, or after a period of time. For this reason, it is often worth while to insert triangular fillets into the formwork at such points.

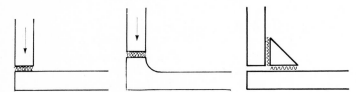

Butt-joints in the angles of formwork.

Heavier chamfering is (in the author's opinion) unnecessary, since it doubles the problem of making the joints grout-tight. Right-angle corners are easier to form. If properly prepared, they will in practice prove quite satisfactory.

Visual concrete requires substantially more stable formwork than does ordinary concrete, because all irregularities and indentations in the finished surface will remain permanently visible. Moreover, the process of compacting by vibrator imposes considerable strain on the formwork.

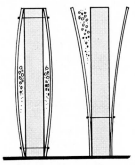

The consequences of over-economical shuttering.

Spacers and tie-wires
The formwork for low concrete walls can be stiffened by means of wedged struts along its sides.

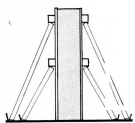

Bracing a low wall by means of struts.

In higher walls, opposing faces of formwork need to be held together by *wire ties*, which are nowadays generally threaded through the concrete in plastic pipes. It is true that because of these ties the poker-vibrator must be more often withdrawn and re-inserted; but the increased pressure-resistance which ties give to the formwork is essential.

The purpose of ties is to maintain an even wall thickness. Best for the purpose are bolts or steel wires threaded through plastic pipes. This achieves the combination of a pressure-resistant wall of the planned thickness with great stability in both surfaces of the formwork. After striking, the wire or steel ties can be easily withdrawn from the finished concrete.

Above: A spacer made of hard PVC which has only point-contact with the formwork.
Below: A spacer of PVC tubing with a tie rod threaded through it.

The spacing between formwork and the outermost rods of reinforcement is best maintained with the aid of plastic *spacers* which come into contact with the formwork only at isolated points. If such spacers are used, however, care must be taken in steam-curing, for unsuitable plastic will melt.

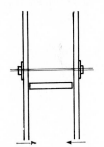 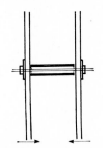

The usual method of bracing, with spacers and ties separate. Not recommended.

Spacer and tie combined. This is better practice.

The use of spacers of timber cut across the grain, though it is cheap, is to be discouraged, for such spacers leave marks on the concrete.

Tie-holes, even when carefully made good, always remain to some extent visible. Since they can often affect the appearance of a building, they should always be approved by the architect in collaboration with the structural engineer or the contractors. If concealed in a shadow groove, tie-holes can be made practically unnoticeable.

Lift joints and access panels

Since lifts of full wall height are not usual in concreting, there is always the danger that the joins between lifts cast on different days may stand out too visibly. If shadow grooves can be planned to occur at points where concreting for the day is likely to finish, such irregularities in the concrete structure can be made less obvious. Such grooves, moreover, will act as useful contraction joints at the dangerous points where lifts cast at different times meet.

For the better control of compaction in difficult places, it is sometimes worth providing *access panels* in the formwork, which can also be used for insertion of the vibrator.

The drawings for the formwork should indicate not only the method and pattern of construction, but also the position of ties and construction joints. The positioning of the latter must take into account the working schedule, and so involves consultation with the structural engineer and the contractors.

The construction of good grout-tight formwork requires precision carpentry falling not far short of the standard of the cabinet-maker.

Special importance attaches to tolerances in the construction of formwork because of the necessity for connecting further components (plumbing, electrical wiring, etc.), or additional buildings. In Germany, maximum permitted deviation from the dimensions laid down is governed by DIN 18201.

Reinforcement

The reinforcement of buildings in visual concrete requires special attention. The concrete cover must not only be sufficiently thick (35 mm), but must also not betray the presence of the reinforcement to an observer. Closely packed horizontal or inclined reinforcement makes uniform placing of the concrete difficult. Given aggregate of a maximum size of 30 mm ($1\frac{1}{4}''$), the rods should be placed vertically and at least 35 mm ($1\frac{1}{2}''$) from the formwork so as to allow unrestricted settling of the concrete and so prevent the formation of cavities. In this way the resistance of reinforced concrete buildings to frost, and particularly the resistance of the reinforcement itself to fire, is much increased.

The reinforcement should be lowered into the formwork after it has been assembled, and the rods should be braced and tied against one another and against the formwork in such a way that vibration cannot move them out of position.

For successful compaction, allowance must be made in the design of the reinforcement for inserting the vibrator. This is especially important where slender columns with a high finish are specified.

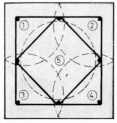

Channels (1–5) for insertion of the poker vibrator.

Heavily reinforced lintels over doors and windows are very difficult to finish in exposed concrete. One possible method is to place the construction joint immediately under the reinforcement, or to arrange for vibration to be applied through an access panel lying under the reinforcement.

The blemishes in the concrete round the door opening are a consequence of insufficiently watertight joints in the formwork.

In the U.K., the usual practice is to employ external vibration as a means of improving the compaction of reinforced-concrete lintels.

Placing the concrete

It has been said that the formwork must be so carefully planned and erected that the entire frame will be sufficiently strong to withstand the pressure of the liquid concrete and any risk of deformation during its subsequent working.

Horizontal planes such as the soffits to lintels and floors should be shuttered with a camber, since this is optically less disturbing than is a sagging line.

Continuous work with the plumb-line and water-level are needed to ensure accuracy in the erection of formwork. Levelling with the theodolite between storeys is recommended. The condition of the formwork should be continuously checked during the placing operation itself, and if necessary corrected by means of wedging, extra struts etc. Wedging can also help if the formwork proves not to be watertight. Alternatively, leaking joints can be sealed with wooden chips or putty—but never with plaster.

The formwork *must* be made watertight, for any grout escaping through interstices in it will show up as rough, uneven lines on the surface of visual concrete.

Tight joints in formwork can be achieved by means of tongued and grooved boards; but such boards are often difficult to re-use because they are easily damaged during the operation of striking. Alternatively, self-adhesive foam-rubber strips can be inserted into the joints; or the joints can be filled with a synthetic mastic which will not harden when exposed to the air.

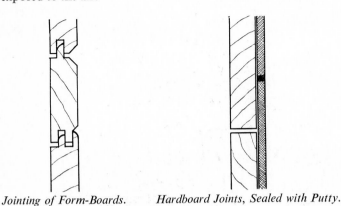

Jointing of Form-Boards.   Hardboard Joints, Sealed with Putty.

Surface treatment of formwork
Oiled or waxed boards are struck more easily and are easier to keep clean than are untreated ones. There is always the danger that parts of a board will soak up too much mould oil and that a darker patch will then show up on the concrete surface.

Formwork panels must be treated in accordance with the instructions of the suppliers if the appearance of spots of mould oil on the surface of the concrete is to be prevented. Laminated panels, especially ones built up of different materials, have been known to blister when exposed to strong sunlight. Joints between large areas of panel can be sealed with strips of foam rubber or with mastic.

For smooth visual-concrete surfaces, all nails should be countersunk into the formwork and puttied over.

Jointing of formwork
It is important to ensure the correct sealing of the formwork when extra shuttering for the pouring of more concrete is being added. Great care must be taken to see that the new formwork is securely clamped to the concrete sections already placed. Foam-rubber strips to which the new formwork is tightly pressed can be used.

Mixing the concrete
To avoid later argument, it is essential to prepare samples of the concrete mix to be used, before work begins. The composition and appearance of the sample finally chosen, including all admixtures and pigments, should remain *mandatory* throughout the work of construction.

Quantitative analysis of the agreed mix should be determined by weigh-batching and water-meter only. Really accurate measurement by weight is an essential precondition to the making of good visual concrete. It is even more important than are constant mixing times.

During concreting, continuous control of aggregate size and grading, and of the consistency and weight of the new mixed concrete, is essential. Control notes should always be recorded.

The quantity of concrete made per mix should be sufficient for a lift of 20–30 cm (8–12″) throughout. Meal-breaks should not be allowed to interfere with the completion of a lift, for otherwise the joints will show up later on. After an interruption of more than half-an-hour, the lower strata of the lift can no longer be vibrated, the chemical process of setting having by then progressed too far. Breaks of any kind should therefore be permitted only when a point has been reached at which the architect has made provision for an architectural feature designed to make a working joint invisible.

Placing the concrete
Concrete should be placed through funnels spaced as closely as possible so as to avoid the forming of cones on whose rims uneven distribution of the aggregate can take place. The maximum height of free fall should be not more than 4 m (13 ft) lest the mix segregate and air get into the concrete. To avoid high lifts, the inner formwork is sometimes raised as the work proceeds; but this tends to be costly on account of the need for continual coordination of formwork and concreting. Varying heights of fall result in different degrees of compaction in the placed concrete.

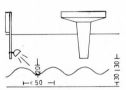

Short Intervals: Low Placing Height: Handlamp to be used where necessary.

Funnels and pipes used for casting should not be emptied by using the vibrator. A hand rod should be used instead. Otherwise the concrete will pre-compact, with resulting unevennesses on its face.

All tall, slender members need to be filled with special care. Control can be improved by the use of hand-lamps, and sometimes also with the aid of observation panels.

Lift joints and work joints
If fresh concrete is placed on top of unhardened concrete and the previous layer can be re-vibrated together with the new one, lift joints will show less.

To make a work joint practically invisible in a smooth wall, the procedure is as follows: A planed lath is fitted to the formwork in such a way that its lower edge coincides precisely with a shutter joint, after which concreting is continued up to the upper edge of the lath. When the lath (whose interior surface should be bevelled) is removed, an exact edge will have been formed which will be made invisible, as concreting proceeds, by the shutter joint. It is essential that care be taken to make the joint grout-tight.

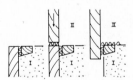

(a) *Lift 1 concreting completed. A smooth finish is obtained by means of a batten inset.*
(b) *First Variation: Formwork continued by means of filled butt joints.*
(c) *Second Variation: Formwork struck; new formwork erected as shown; joint invisible.*

Concreting in cold weather
The colder the concrete, the longer it will take to set and harden. At 5 °C the process of setting comes more or less to a stop. Ultimately, the water in the mix freezes and breaks up the hardening concrete. If concrete freezes during curing, its strength will suffer; and under certain circumstances it will become completely useless. Concrete which has frozen can be recognized later on by the fern-like impressions left on its

Unwrought shuttering coarsely jointed.

Shuttering planed, tongued and grooved.

(Photographs by Helmut Pörtner, Wiesbaden)

Steel formwork.

Plywood formwork faced with plastic.

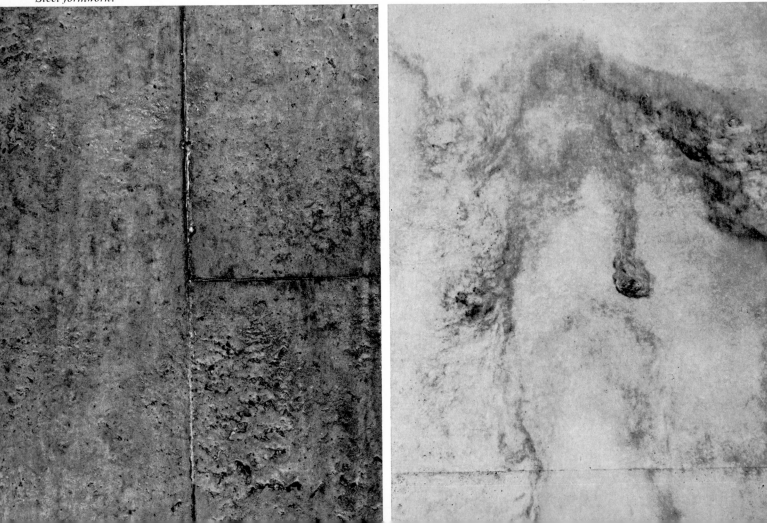

surface by the ice. Concrete is safe from frost only when it has thoroughly set.

At temperatures below 5 °C, concreting should only take place with special precautions. The concrete must have, and must retain, the heat necessary for the curing process. The mix should never be colder than 5 °C. Frozen aggregates must never be used, and the gravel/sand and water should be heated before use. (*The cement itself should never be heated.*)

Newly-concreted structures need to be protected from heat loss by covering them with straw, sawdust, boards or several layers of paper; but care must be taken to prevent staining of the concrete by tannin from the wood, straw or sawdust used for protection. Steel shuttering abstracts heat from concrete more quickly than does wooden formwork. Anti-freeze admixtures are available which speed up the process of hydration, and in doing so generate sufficient heat to allow concreting to be carried out at temperatures as low as −10 °C.

Experts on the subject should, however, be consulted, for some accelerating admixtures such as calcium chloride can cause corrosion of the reinforcement. (*Note by British Editor— Many authorities consider that, for this reason, calcium chloride should never be used in the making of exposed concrete.*)

Compaction
Vibration is used to make relatively stiff mixes more fluid. In this way a watertight concrete can be attained with even a low water/cement ratio. The correct frequency and amplitude of vibration, and the correct vibration periods, should be ascertained by experiment.

The vibrator should be plunged into the mix quickly and withdrawn more slowly, the correct ratio of withdrawing to plunging speed being of the order of 3 : 1 to 5 : 1. Only so can air bubbles escape easily and the concrete flow smoothly together behind the vibrator.

The minimum time to allow for vibration is 20 minutes per cubic metre of concrete if good visual concrete is to be produced. (This means that the vibration period must be considerably longer than that laid down in the German Standard DIN 4235.) Even with a water/cement ratio higher than 0·45, there is no risk that the mix will segregate through vibration if one keeps to the prescribed vibration times. On surfaces having an inward slope vibration must be carried out with especial care.

Compacting the concrete after the curing process has started is not only permissible but can be positively beneficial—for a number of reasons, including the improvement of lift joints and of the quality of the lower layers of concrete. For this reason the vibrator should occasionally be plunged a distance of 10 to 30 cm (4–12″) into the previous lift.

Vibrators working on the outside of the formwork and on the same level as vibrators working inside can further improve the concrete by helping the escape of air bubbles. External vibration alone, however, is not sufficient for the production of good visual concrete, though it can be helpful in the compaction of members too slender to allow the vibrator to be inserted into them.

Pre-fabricated components of a concrete structure are normally compacted on vibrating tables. If this is done carefully, air pockets and honey-combing will be rare.

Dampening of the formwork
To prevent the formwork from warping and/or leaking, it should be kept evenly damp from before the commencement of placing right through to completion of the curing process in the concrete. In the full glare of the sun, spraying is not sufficient. The formwork must be kept in the shade by curtaining it with hessian drapes.

Constant checking of the formwork is essential, from the time it is completed to the time it is to be struck.

Grout oozing from open joints and other points can cause premature drying and colour variations at the escape points. Any grout seen to be running down over visual concrete must be hosed and brushed off immediately.

Protection of exposed reinforcement
Steel reinforcement left exposed to the weather in winter, or for long periods prior to being used as jointing for future extensions, will tend to rust; and traces of rust stain are difficult to remove from concrete surfaces. To avoid this danger, all exposed reinforcement should be coated with cement grout or wrapped in plastic sheet for protection.

Striking of the formwork
This operation should seldom be begun earlier than seven days after placing. The concrete will in this way receive longer protection against too-rapid cooling, and the danger of shrinkage cracks will be much reduced. The striking of the form-work over visual-concrete surfaces demands special care—and it should be a rule that all crowbars are locked away before work begins!

It is for the architect to decide whether the concrete ridges corresponding to joints in the shuttering should remain, or whether they should be cut away immediately after the form-work is struck.

When striking is complete, the wire ties are pulled out of the plastic pipes and the holes sealed off with PVC mortar or plastic pellets. Ties which cannot be pulled out because they have been jammed in by the vibrator or for any other reason should be cut back at least 20 mm and preferably 30 mm (say, about an inch) below the concrete surface. The hole must then be carefully plugged to prevent rust stain coming through.

Ideal would be a tie of rust-proof material, available at an economic price, which could be cut off at the surface of the concrete and remain there without deleterious effect.

After-treatment of the placed concrete
Fresh concrete needs to be properly cured. In particular, it must be kept damp, if possible for 21 days; for uneven drying will otherwise create surface tension. But the dampening is better accomplished by covering or wrapping the exposed concrete in plastic sheeting, rather than by spraying. There are available membranes which can be applied to slow down the drying-out of the surplus water, but they always affect the colour of the concrete and are for that reason seldom suitable for use on visual concrete.

Never cool a cast-concrete structure too rapidly. In no circumstances must cold water be used to hose down freshly-exposed concrete. At this point in time, the wall will be heated internally by the energy released in the curing process to a temperature of 45 °C (even 85 °C in thick walls), and abrupt cooling with cold water will cause surface cracks.

Protecting visual concrete against soiling and damage
Freshly-exposed concrete soon picks up dirt on a building site. A single coat of water-repellent applied immediately after the curing process is complete allows any dirt which may adhere to be easily cleaned off.

All surfaces exposed to damage during building operations, especially external angles, arrises and the like, should be protected by being boarded up, using a soft underlay. The plastic sheeting used to assist the curing process itself gives some protection. Since they have proved successful in practice, it is recommended that plastic coverings of this type be included in the specification itself.

Cleaning

For washing down concrete surfaces, clean water, or a solution containing a common household detergent, is often sufficient.

Effective protection of smooth concrete surfaces against pencil, ball-point pen and oil crayon marks is provided by applying an 8% solution of oxalic acid. Against Indian-ink pencil, a 10% solution of phosphoric acid is effective. Rough surfaces present a more difficult problem; but pencil and crayon marks can be removed with trichlorethylene.

Rust stains can be removed by brushing with a solution of one part sodium citrate to six parts water, and by dusting afterwards with sodium-hyposulphite crystals.

To remove black or green discoloration caused by copper or bronze, apply a paste of one part ammonium chloride to four parts of a suitable filler such as diatomaceous earth, mixed with spirit of ammonium chloride. Allow to dry, then brush off —repeating the process if necessary.

Before treatment begins, all surfaces to be cleaned must be thoroughly wetted with clean water. All cleaning agents used must then be completely washed off, neutralized and dried. If this is not done, remnants of acid and other chemicals may themselves cause spots and efflorescence.

Patching

Any faults which may appear in visual concrete should, in the author's opinion, be left untreated. Patches can never in practice be completely hidden; and visual concrete seldom allows itself to be made completely perfect.

But any exposed reinforcement must, without fail, be protected by making good with mortar. The same treatment should be applied to areas of serious honeycombing which could allow frost to cause damage.

It is very difficult to give the appropriate colouring to mortar used for patching; but there are plastic emulsions on the market which assist the adhesion of new concrete to old.

Water-repellent treatment

Concrete absorbs moisture more or less easily depending on its density and on the suctional efficiency of its capillary structure. Dust deposited by rain settles in the course of time in the minute pores, and soils the surface. Although rough surfaces in fact get dirtier than do smooth ones, the dirt shows up more on the latter if they are not treated with water-repellent.

The application of a silicone-based water-repellent slows down the soiling process. Application is by spraying—the character and colour of the surface being affected hardly at all. The solution does not hinder the evaporation of water-vapour, but it does prevent the penetration of damp and the consequent danger of hair-cracks caused by the expansion and contraction of the surface concrete.

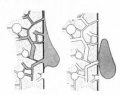

The effect of treating with a water-repellent.

Interior walls especially exposed to soiling can be made washable by the application of colourless plastic emulsions. Concrete floors can be sealed with a synthetic resin finish, or treated with a silicate-based hardener which helps to bind the surface and to give greater resistance to wear.

Washed concrete (exposed-aggregate concrete)

If the formwork be struck relatively early, the not-yet-quite-hardened outer skin of the cast concrete can be removed by washing and brushing so as to expose the texture of the larger constituents of the aggregate, which are usually gravel. If the concrete has hardened, a diluted solution of hydrochloric acid (about 10%) can be used to achieve the same effect; but it must then be carefully neutralized and washed off.

To achieve the best appearance by this method, careful selection and grading of the aggregate is essential. Sampling tests are strongly recommended.

By applying to the inside of the formwork a retarder which slows down the process of hardening (sugar was formerly used for the purpose), it is possible to obtain more time for this washing treatment. The correct intervals between casting, striking and washing should be ascertained from the suppliers and their advisers. They depend partly on the quality of the concrete itself, partly on the external temperature. Tests are therefore advisable. On average, the interval between the beginning of the hardening process and the washing will vary between three and 24 hours. The washing process should not be allowed to go too deep lest the aggregate be loosened.

For physical reasons, this treatment cannot be applied to the larger structures cast *in-situ*, because their walls will not have attained sufficient rigidity to support their own weight without formwork. The treatment is therefore more suitable for retaining walls, and is especially successful with pre-fabricated elements whose manufacture can take place in the horizontal position.

Sand-blasting

This is a process which removes the outer skin of the concrete and so exposes the texture of the aggregate-sand combination beneath. The sand-blasting technique can be applied both to large surface areas and to pre-fabricated elements. As with other exposed-aggregate techniques, the cover over the reinforcement should be increased to allow for the material which will be lost.

Tooling of concrete

If concrete be looked upon as a form of artificial stone, the surface techniques of stone-masonry such as bush-hammering, point-tooling or chiselling can be quite satisfactorily applied to it (though there will always be people who suspect that the real reason why a surface has been so treated is that behind it there lies a structure in visual concrete which went wrong!). In any case, the application of such techniques will always be restrained by the cost of the labour involved.

In cases where stone-masonry treatment is specified, the concrete cover must be made appropriately thick to allow for the depth of mechanical tooling. It is not possible for a stone-mason to improve faulty visual concrete surfaces if this adequately thicker cover is not present; for otherwise exposed wire ties, spacers, cramps and the like will disturb the final effect.

Pebbles encountered during handworking should never be gouged out, but should be split instead.

Concrete will not be hard enough to withstand any surface-tooling technique until it is three to four weeks old. Its condition should first be tested at some relatively inconspicuous point.

Polishing concrete

The technique of grinding and polishing large areas of exposed concrete brings out the colour of the pebble to great advantage, but it is not easy to apply on the site. The preparation of ground and polished concrete components in the factory is considerably easier. For this treatment also, the cover to the reinforcement must be made suitably thick.

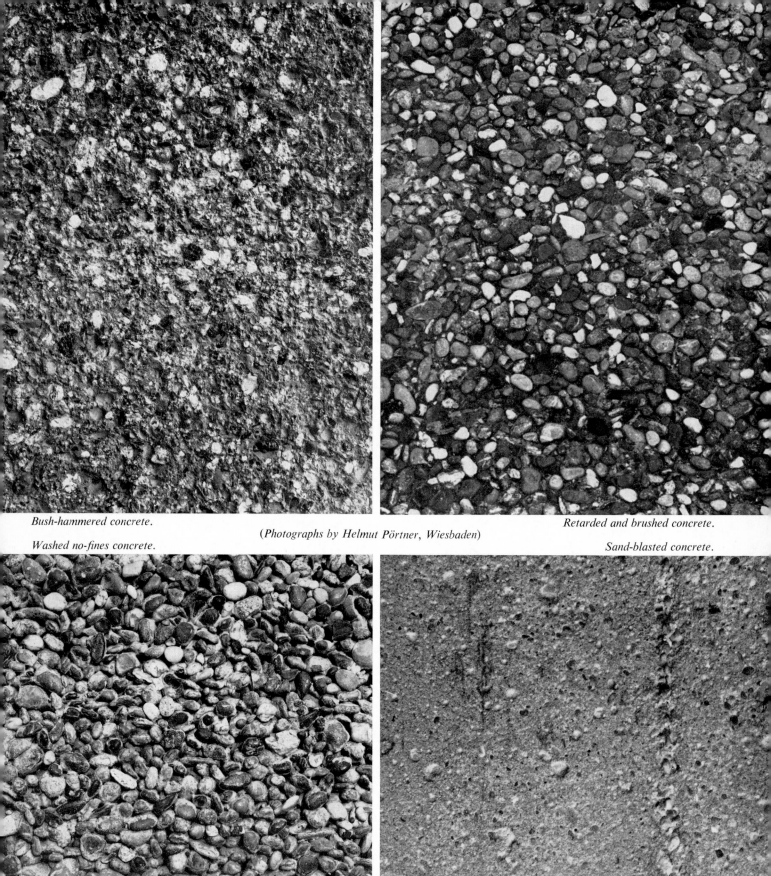

Bush-hammered concrete.

Retarded and brushed concrete.

(Photographs by Helmut Pörtner, Wiesbaden)

Washed no-fines concrete.

Sand-blasted concrete.

Concrete mixes

In Great Britain, standardized concrete mixes tend to be little used, the preference being for designing and testing a special mix for each individual project. Certain recommended maximum and minimum cement contents for mixes to give good visual appearance have, however, been compiled and are set out in the two Tables (in metric and Imperial measures respectively) below:

Maximum size of aggregate	Cement content	Aggregate/cement ratio	Slump*	Sand content† (% by weight of dry aggregate)
Millimetres	*Kg/m³*		*Millimetres*	
37·5	415 to 325		75 ± 20	
20·0	475 to 355	As below	50 ± 10	As below
10·0	535 to 415		25 ± 10	
Inches	*Lb/yard³*		*Inches*	
1½	700 to 550	4 : 1 to 6 : 1	3 ± ¾	35
¾	800 to 600	3·5 : 1 to 5 : 1	2 ± ½	40
⅜	900 to 700	3 : 1 to 4 : 1	1 ± ½	50

* **Slump** is the measure of the workability of concrete used in the U.K.
† The figures are given for Zone 2 sand. They may require adjustment if sand of a different grading is used.

A number of mixes which have proved successful *in German practice* are listed below:

MIX A: B 225—Grey
Aggregates: % of Aggregates (Rhine Gravel) **by weight**
0– 3 mm = 28%
3– 7 mm = 23%
7–15 mm = 20%
15–30 mm = 26%
Powder: 0–0·2 mm = 3% (Quartz)
Cement: PZ 275 = 310 kg/m³
Admixture: Aerating plasticizer
Water/cement ratio: = 0·52
Compacting:
By internal vibrator

MIX B: B 300—Grey:
Hand-compacted
Aggregates: % of Aggregates (Rhine Gravel) **by weight**
0– 3 mm = 25·5%
3– 7 mm = 22·5%
7–30 mm = 52·0%
Powder: 0–0·2 mm = 0·5% (Limestone)
Cement: PZ 375 = 310 kg/m³
Admixture: None
Water/cement ratio: = 0·56
Compacting:
Hand-compacting by punning (timber laths with sheet-steel covering)

MIX C: B 225—Grey
(Gap-graded)
Aggregates: % of Aggregates (Rhine Gravel) **by weight**
0– 3 mm = 39·8%
3– 7 mm = 30·9%
7–15 mm = none
15–30 mm = 29·4%
Powder: None
Cement: PZ 275 = 290 kg/m³
Admixture: Water-reducing agent
Water/cement ratio: =0·64
Compacting:
By internal vibrator

MIX D: B 225—Grey
Aggregates: % of Aggregates (Rhine Gravel) **by weight**
0– 3 mm = 44%
3– 7 mm = 16%
7–15 mm = 20%
15–30 mm = 20%
Powder: None
Cement: PZ 275 = 290 kg/m³
Admixture: Water-reducing agent
Water/cement ratio: Not known
Compacting:
By internal vibrator

MIX E: B 300—Grey
Aggregates: % of Aggregates (Rhine Gravel) **by weight**
0– 3 mm = 30·0%
3– 7 mm = 21·0%
7–15 mm = 24·5%
15–30 mm = 24·5%
Powder: 0–0·2 mm = 4% (Quartz)
Cement: PZ 275 = 300 kg/m³
Admixture: Waterproofer
Water/cement ratio: Not known
Compacting:
By internal vibrator with simultaneous external vibration

MIX F: Blast-furnace cement
B 300—Grey
Aggregates: % of Aggregates (Rhine Gravel) **by weight**
0– 3 mm = 23·5%
3– 7 mm = 25·5%
7–15 mm = 25·5%
15–30 mm = 25·5%
Powder: 0–0·2 mm = 4% (Quartz)
Cement: HOZ 275 = 300 kg/m³
Admixture: Waterproofer
Water/cement ratio: Not known
Compacting: As for Mix E

MIX G: B 450—Grey
Aggregates: % of Aggregates (Rhine Gravel) **by weight**
0– 3 mm = 30%
3– 7 mm = 18%
7–15 mm = 28%
15–30 mm = 24%
Powder: None
Cement: PZ 275 = 340 kg/m³
Admixture: Water-reducing agent
Water/cement ratio: = 0·51
Compacting:
By internal vibrator

MIX H: B 450—White
Aggregates: % of Aggregates (Rhine Gravel) **by weight**
0– 3 mm = 30·0%
3– 7 mm = 21·0%
7–15 mm = 24·5%
15–30 mm = 24·5%
Powder: 0–0·2 = 3% (Finest white quartz)
Cement: PZ 275 Dyckerhoff-Weiss* = 350 kg/m³
Admixture: Water-reducing agent
Water/Cement ratio: = 0·48
Compacting: As for Mix E

* **Mix H** can be made still brighter by adding to the white cement 2% by weight of a titanium dioxide white pigment. In Britain, up to 5% by weight of titanium dioxide is used in making the whitest grade of concrete.

Some relative properties of building materials
(As determined by German test results)

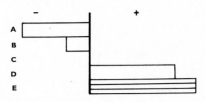

Sound insulation, in dB, of 24 cm layers

A	Cellular concrete (GS 50)	− 1
B	Pumice stone (solid) (V 25)	− 3
C	Perforated brick (HLz 1·2/100)	± 0
D	Wall brick (MZ 100)	+ 4
E	Concrete (B > 225)	+ 5

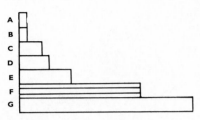

Thermal conductivity, in kcal/m × °C

A	Woodwool/cement panels (*Heraklith*)	0·12
B	Pinewood	0·12
C	Pumice stone cavity block (Zwk, Hbl 25)	0·30
D	Cellular concrete (GS 25)	0·42
E	Wall brick (MZ 100)	0·68
F	Concrete (B > 160)	1·75
G	Natural stone	*about* 2·50

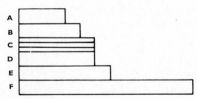

Coefficients of thermal expansion, in cm/cm × °C

A	Wall brick (MZ 100)	6×10^{-6}
B	Natural stone	*about* 8×10^{-6}
C	Concrete (B > 225)	10×10^{-6}
D	Asbestos cement	10×10^{-6}
E	Steel	12×10^{-6}
F	Aluminium	23×10^{-6}

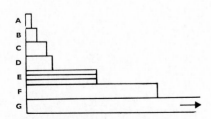

Resistance to absorption of steam (factors)

A	Pumice stone (solid) (V 25)	2·5
B	Cellular concrete (GS 25)	5·4
C	Wall brick (MZ 100)	9·3
D	Lime rendering	11·5
E	Concrete (B > 225)	28·4
F	Asbestos cement	60
G	Bituminous paperboard, 500 gr/m²	1300

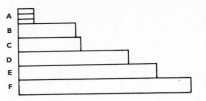

Saturation-point humidity (water absorbed as % of weight)

A	Concrete (B 300)	5
B	Lime rendering	19
C	Wall brick (MZ 100)	20
D	Gypsum rendering	35
E	Pumice stone cavity block (Hbl 25)	44
F	Cellular concrete (GS 25)	55

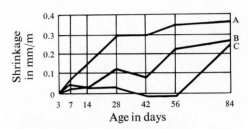

Concrete shrinkage in different conditions of storage
(Source: Research and Materials-Testing Institute, Stuttgart)
Curve A = Stored at 12–24°C and at 50–60% of air humidity.
Curve B = Stored for 42 days in polythene foil: then as in A.
Curve C = Stored for 56 days in damp box at 20°C and 95% relative humidity: then at 20°C and 65% relative humidity.

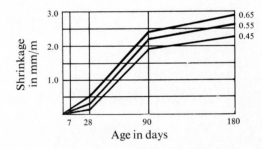

The effect of extra water on concrete shrinkage
The figures to the right of the scale indicate the water/cement ratio

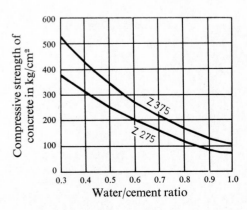

**The effect of additional water on the strength of concrete
(Cements Z 275 and Z 375)**

15

Chapter 2

Monolithic structures in visual concrete

In the Beginning . . .

"Architecture in large cities like Paris, where the slightest whim of the Emperor causes whole districts to be demolished and new ones to arise, is required nowadays to fulfil these demands in an incredibly short space of time. To meet them, a new, cheap and easily produced material has had to be found. It has been discovered in the form of a material already available but not yet used to any great extent, namely: concrete.

"Concrete is a crude mortar prepared from $1\frac{1}{2}$ parts of river gravel, $2\frac{1}{2}$ parts of sharp sand, $1\frac{1}{2}$ parts of lime and $1\frac{1}{2}$ parts of boiling water. With this material the construction of houses is not much more difficult than is the making of cheese. There is little difference between the methods of the pastry cook and those of the mason when the houses need only to be cast into a mould. . . ."

From the magazine *Über Land und Meer*, No. 7, 1859

Unity Temple in Oak Park, Illinois (1906)

Architect: Frank Lloyd Wright

(*Photograph by Chicago Architectural Photo Co.*)

An early example of visual concrete used in architecture. The outlines of the shuttering used in the casting process are clearly visible in the photograph.

Unfortunately, the building has now been rendered.

Frank Lloyd Wright *wrote in his celebrated* Testament *that, so far as he knew, this was the first time a ready-cast block of concrete had been used as a building component. The block was cast in wooden forms and wooden boxes, he wrote, and bore the marks of this technique.*
The first step had been the destruction of the box itself, from which there had resulted an impression of the box's interior space as the essence of the whole building.

The Shinseisaku Theatre, Cultural Centre, Tokyo (1964)
Architect: Research Institute of Architecture, Tokyo
(*Photograph by Masao Arai, "The Japan Architect"*)

In the open-air theatre, the huge rear wall of the building
serves as backdrop to the stage. Tooled surfaces were
superimposed on the pattern of the formwork.

The Cultural Centre of Nichinan (1962)

Architect: Kenzo Tange, Tokyo

(*Photograph by Retoria/ Y. Futagawa, Tokyo*)

The Cultural Centre is intended to form, in conjunction with the Town Hall, the centre of a new city which is growing out of three fusing localities.

Whereas in the Shinseisaku Theatre (*see opposite*) the great expanse of the rear wall is used to form a deliberately anonymous background to the stage, here the structure gains tension and meaning from the contrapuntal distribution of openings and moulded components. The severely formal aspect of the building matches its purpose.

19

**The Protestant Church on the Blumenau, Mannheim,
West Germany (1960–1961)**

Architect: Helmut Striffler, Mannheim Lindenhof
Associate: Franz Katzenberger

Scale of section, 1:200

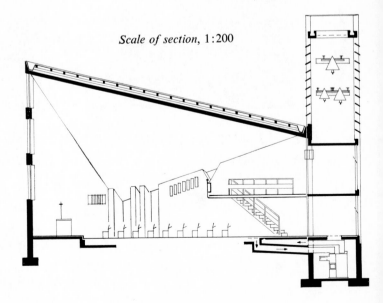

The Protestant Church on the Blumenau

(Photographs by Robert Häusser, Mannheim-Käfertal)

Exterior wall, showing expansion joint

The Pulpit

Concrete mix: Grey Portland cement, 300 kg per cubic metre. Aggregates supplied separately to site. No extra-fine aggregate added. Walls untreated outside; silicone fluate applied inside.

Formwork of boards about 10 cm wide, planed on three sides but with the unplaned face towards the concrete. Butt joints. Longitudinal joints staggered.

Wall and roof concreted in a single operation to avoid working joints and to allow for variations of shrinkage in different parts of the structure. Although the bulk of the concrete was cast at almost the same time, a vertical expansion joint was provided to protect the walls. (It is clearly visible running down from the triangular window in the photograph above.)

Exterior wall of 30 cm reinforced concrete; no heat insulation, warmed-air heating being installed in the interior. Windows glazed with 35 mm rough-cast plate glass, puttied unframed into grooves in the concrete.

The Protestant Church of the Reconciliation at Mannheim-Rheinau, West Germany (1963–1965)

Architect: Helmut Striffler, Mannheim-Lindenhof
Associate: Detlef Beron

(*Photograph by Robert Häusser, Mannheim-Käfertal*)

Concrete mix: 300 kg grey Portland cement and 60 kg Trass cement per cubic metre. Formwork of boards some 10 cm wide, planed on three sides but with the fourth (unplaned) side towards the concrete. Tongued and grooved. The usual mould oil.

Walls 60 cm thick, with the concrete exposed on both sides. Insulation by means of 3 cm-thick expanded polystyrene layer in the wall itself, protected on both sides by metal mesh. Both internal and external skins were cast simultaneously in a single lift. No damp course. Walls untreated on the outer side, but silicone fluate applied on the inner side.

The casting joints were agreed at the design stage with the contractor and structural engineer, and were formed by battens of dove-tail section. The design aimed at making possible concrete lifts of minimum width and maximum height. The greatest breadth to be poured in one lift was 4 metres, the greatest height 5 metres. Both skins were placed with the aid of tremie pipes to achieve a constantly even pressure on the insulating core. The formwork was struck after 8–10 days—certainly no earlier.

The Protestant Church at Mannheim-Rheinau
View from the East

All exposed arrises were protected by boarding. Plastic sheets protected the main walls themselves from spilt grout after concreting, and the walls were also hosed down immediately after pouring

One important piece of experience gained was that any steel reinforcement left exposed during extended working breaks caused by bad weather needs to be protected against rain. Rust stains running down the concrete face are practically indelible.

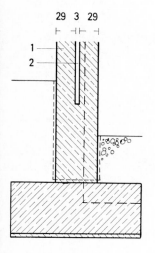

Section of Wall on a scale 1 : 50
1. Exposed concrete.
2. Expanded polystyrene layer, 30 mm thick, protected on both sides by metal mesh.

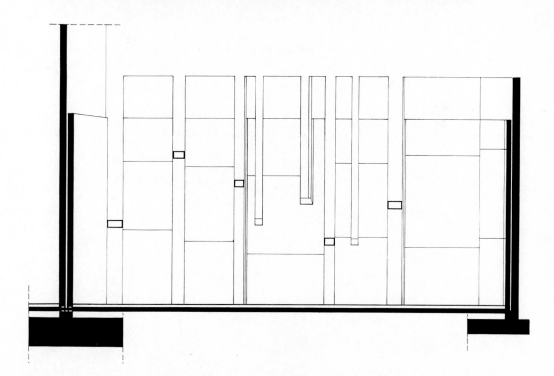

South Wall: Interior Elevation (*above*); Section (*below*)

Scale, 1 : 200

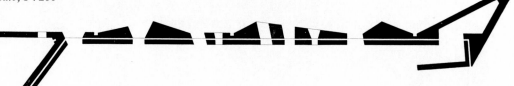

23

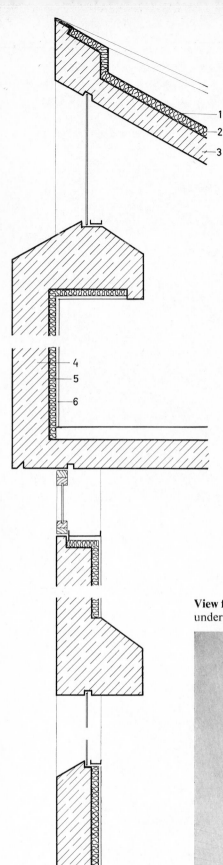

The Community Centre in Frankfurt-Sindlingen, West Germany (1961–1962)

Architect: Günter Bock, Frankfurt/Main

(*Photographs by Jupp Falke, Frankfurt/Main*)

Built as the community centre of a garden city dating from the 1920's, the Sindlingen complex contains an assembly hall seating 550 people, a restaurant, a library, club rooms and workshops for young people.

Constructed of horizontally-staggered wall slabs of reinforced concrete with V-shaped reinforced-concrete roof units. Finished with silicone paint.

The formwork was assembled of the usual shutter panels of varying sizes, to conform to an overall formwork design. Construction joints and panel butt joints emphasized by battens attached to formwork. No mould oil.

Insulation by means of 3·5 cm-thick light-weight woodwool boards fixed to the inside of the formwork. All inner wall faces rendered. The assembly hall has a plasterboard skin with 1 cm of air space behind.

The glass was in most cases puttied directly into prepared grooves in the concrete. To cast these grooves with the required accuracy was not easy.

In the section drawings on the left,
the scale is 1 : 20
Key to the small numerals:
1. Three-layer felt roof.
2. Cork 4 cm thick.
3. Roof slab of reinforced concrete.
4. Main walls of reinforced concrete 21·5 cm thick.
5. Lightweight woodwool boards 3·5 cm thick.
6. Plaster 1·5 cm thick.

View from the West. The main entrance to the Assembly Hall is under the projecting wing on the left.

The Frankfurt-Sindlingen Community Centre

The South-east face showing emergency exits from the Assembly Hall.

Note the roof drainage pipes set well out from the façade.

Connection of rainwater heads and down-pipes.

Section and elevation to scale of 1 : 20
Key:
1. Rainwater head of steel, concreted in, formed of two angles. 160 × 160 × 15 mm.
2. Down-pipe, 10·5 cm.
3. Three-layer felt roof.
4 and 5. Cork, 2 cm thick.

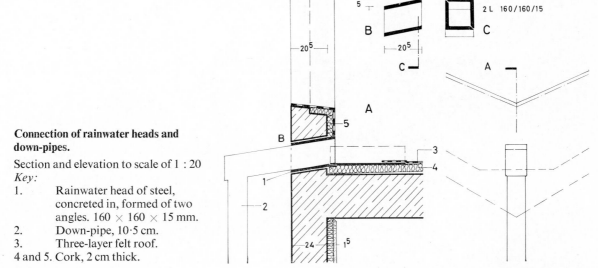

The Setagaya Community Centre, Tokyo (1959)

Architects: Kunio Mayekawa & Associates, Tokyo

(*Photographs by Yoshio Watanabe and Ch. Hirayama, Tokyo*)

The interior (*above*) and exterior of the Centre achieve a monumental effect by unity of form and material.

Concrete mix (per cubic metre): 327 kg Portland cement, 692 kg sand, 1139 kg gravel. Colourless plastic paint. Formwork of wooden panels 60 × 180 cm.

The Setagaya Community Centre
Roof-line and façade.

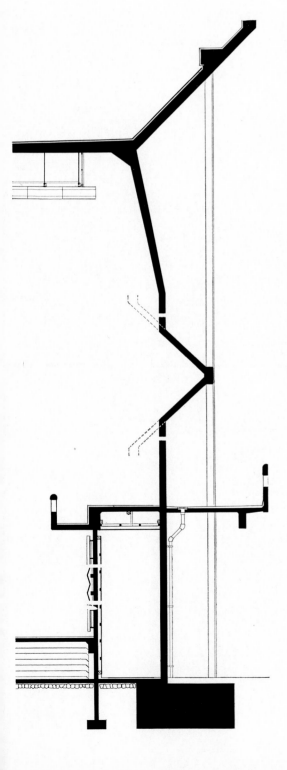

The drawing on the left shows a section of the Assembly Hall on a scale of 1 : 100.

German Fortifications in the Channel Islands during the Second World War

(*All photographs on this and the facing page by Colin Partridge, London*)

A Gun Emplacement at L'Etacq, Jersey

A Coastal Battery Emplacement, St. Ouens Bay, Jersey

A Fire-control Bunker on Mannez Common in the Island of Alderney

An Observation Post at La Moye, Jersey

Coastal Fortifications in the South of France

The bunker shown in the photograph above is at Agde, in
Languedoc.

(*Photographs on this page and the page opposite by Eric de Maré,
London*)

A Second-World-War Bunker on the South Coast of France

Castles once served man for his protection and defence. The castles of today are concrete bunkers and air-raid shelters. Both originated without known author; both lacked either artistic or formal purpose; and both arose untrammeled by inherited formula in their configuration.

In the case illustrated above, architectural pointlessness is married to considerable plastic attraction. An interesting contribution to the theme: "Architecture without architects". . . .

The Protestant Church at Reinach, Canton Basle, Switzerland (1962)

Architect: Ernst Gisel, Zürich

(*Photographs by Fritz Maurer, Zürich*)

Concrete mix B300 (*see Chapter 1*) with grey Portland cement, sand and aggregate grading up to 30 mm. A very dry mix, and no surface treatment.

Formwork of rough boarding 30 mm thick, 12 and 16 cm wide, with board lengths staggered. The boards were re-used three or four times. Joints were made grout-tight with the aid of 4 mm-diameter cotton cord, forced in.

The formwork was sprayed with water for 24 hours before casting so as to prevent any movement in it which could affect the surface texture. (If there is risk of exposure to hot sunshine, the formwork should also be kept damp after concreting has been completed.)

The formwork was struck within 48 hours of placing in order to preserve the texture of the boarding and to allow simple cleaning of the cast surface with brush and water.

Working joints were so arranged that the last board of the previous lift was secured to the first board of the next lift by a screw passing right through both. These screws were tightened before shuttering of the second lift began. In this way all working joints were made leak-proof, and smudging of the concrete was prevented.

Heat insulation with 3 cm of cork. Inside, the walls are finished with 12 cm-thick fair-faced brickwork or with wood panelling.

Note in the photograph to the left how economy of interruption in the façades preserves the solidarity of the building outline. The transverse apertures of the windows intensify the sculptural effect, and underline the structural and plastic possibilities of concrete.

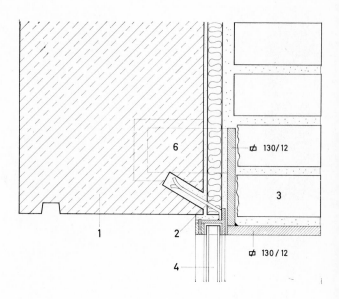

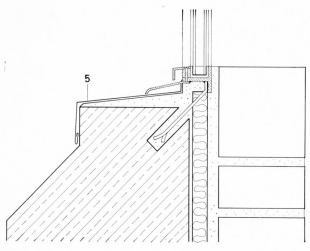

Scale of both sections 1 : 5

Key:
1. Reinforced concrete, 25 cm thick.
2. Cork, 3 cm thick.
3. Brick backing, 12 cm thick.
4. Double glazing.
5. Copper sheeting.
6. Anchoring lug for the lintel.

Two other Views of the
Protestant Church at
Reinach

33

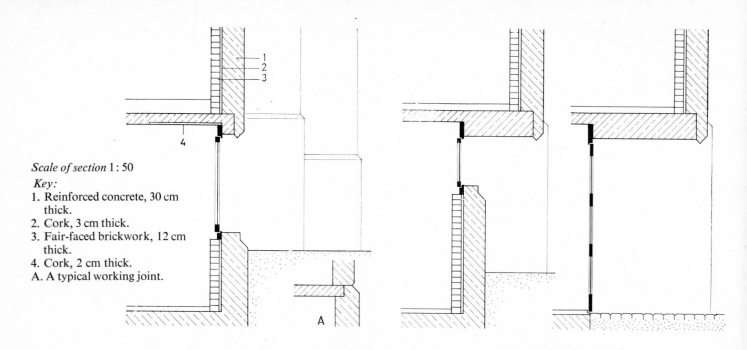

Scale of section 1:50

Key:
1. Reinforced concrete, 30 cm thick.
2. Cork, 3 cm thick.
3. Fair-faced brickwork, 12 cm thick.
4. Cork, 2 cm thick.
A. A typical working joint.

The Reformed Church at Effretikon: Detail of the North façade

Scale of Section 1 : 5
Key:
1. Double-glazing unit.
2. Putty.
3. Copper sill, 2 mm thick.
4. Uppermost layers of brick backing (bricks laid upright).
5. Cork, 3 cm thick.

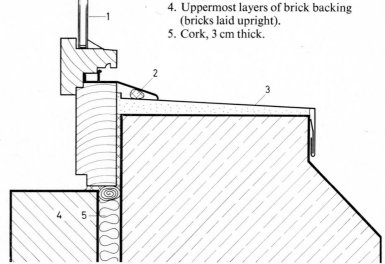

The Reformed Church at Effretikon, Switzerland (1960)

Architect: Ernest Gisel, Zürich
Associate: Louis Plüss

(*Photographs by Max Hellstern, Regensberg*)

Concrete mix: Grey Portland cement, sand and coarse aggregate grading up to 30 mm. Very dry mix. No surface treatment. Formwork of rough boarding 30 mm thick, 12 and 16 cm wide, with board lengths staggered. The boards were re-used three or four times. Joints were made grout-tight with the aid of cotton cord 4 mm in diameter, forced in.

The formwork was sprayed with water for 24 hours before placing of the concrete began so as to prevent any movement which could affect the surface texture. (If there is risk of exposure to hot sunshine, the formwork should also be kept damp after concreting has been completed.)

The formwork was struck within 48 hours of placing in order to preserve the texture of the boarding, and to allow simple cleaning of the cast surface with brush and water.

The whole building forms a harmonious combination of display concrete with black asbestos roof-tiles and copper sheeting which has been much copied.

Secondary School at Aesch, near Basle, Switzerland (1960–1962)

Architects: Walter M. Förderer, Rolf G. Otto and Hans Zwimpfer, Basle and St. Gallen

(*Photographs by Fritz Maurer, Zürich*)

The East playground

Formwork of wrought boards 12 cm wide, chamfered and grooved. Board lengths staggered at least 1 metre.

Internal insulation by means of 5 cm cork faced with wood panelling on battens.

When the concrete was being placed, wet sacking was used to protect the formwork. Spilt laitance was washed off with the hose.

A rainwater spout at window-sill level in the School at Aesch

The Secondary School at Aesch seen from the South-west

In this school, the use of concrete as a building material has been exploited to the uttermost. The dominance of the plastic form is shown both in the overall effect and in detail; but it can only be achieved by pre-planning carried to the ultimate extent.

Concrete here displays to perfection its "monumental" character as a building material.

Terraced Housing Estate at Mühlehalde, Umiken bei Brugg, Canton Aargau, Switzerland (1963–1966)

Architects: Team 2000 (Scherer + Strickler + Weber), Zürich

Outside walls of 15–20 cm reinforced concrete. Heat insulation by means of 5 cm reed matting on battens, rendered, with 2 cm-thick foamed plastic boards fitted behind built-in cupboards. Floors of 16 cm reinforced concrete, with floating screed. Flooring of cleft tiles. Roofing slab insulated with 4 cm cork between damp course and roofing, on top of which was laid a protective screed and 30–50 cm of humus.

The usual insulation of "cold bridges" by means of 50 cm-wide cork margins to roofs and floors.

The overall design was determined by the structural principles, which allowed considerable freedom in the planning of individual units. Work-rooms were fitted in at the final stage. Window openings in the side walls were arranged, within permissible technical limits, to suit individual requirements.

The long horizontal windows are typical, not of this particular project only, but of concrete design in general. Form grows directly out of the material used and its structural properties. It is in such details as these that concrete differs most sharply from brick or stone as a building material.

"The Hanging Gardens of Mühlehalde"

"The architect creates the grand design, the people add their bits and pieces"—*Scherer.*

Below: **Another view of "The Hanging Gardens"**

Hotel on the Godesberg, Bad Godesberg (1961)

Architect: Professor Gottfried Böhm, Köln-Marienburg
(*Photographs by Artur Pfau, Mannheim-Feudenheim*)

The Hotel Entrance

A hotel and restaurant have been built into the old stonework of a ruined castle on the Godesberg above the town of Bad Godesberg on the Rhine.

Concrete of dark blue-grey Portland cement, 200 kg per cubic metre. Water/cement ratio 0·5. Blended oil additive. Concrete mixed with aggregate of washed river gravel up to 30 mm in size, plus 10% of medium-sized aggregate 7–15 mm.

Formwork for all vertical planes: tongued and grooved boards wrought on one side. Width of boards between 9 and 11 cm.

Vertical arrises chamfered by triangular planed fillets of 3 cm × 3 cm section. For horizontal surfaces, shutter panels. Vertical planes point-tooled; soffits sand-blasted.

On the inside of the main walls, 2·5 cm of woodwool used as shuttering, finished with lath and 2 cm of lime plaster.

No special heat insulation necessary, even though the concrete layers extend right through from wall to wall.

The Hotel on the Godesberg

Note how the coursed masonry of the old Castle and the new
hand-finished concrete blend into a convincing unit.

Open-Air Bath, Sports and Recreation Centre, Heuried, near Zürich (1961–1965)

Architects: Hans Litz and Fritz Schwartz, Zürich
Associate: Willi Meier, Zürich
(*Photographs by Fritz Schwartz, Zürich*)

Concrete mix: Grey Portland cement, with sand, aggregate and plasticizer. Walls of 20 cm concrete, 5 cm glass-fibre boards and 10 cm lime-sand bricks, fair-faced. PVC foil used as damp course in the shower-rooms.

Formwork of untreated pine boards of equal width, butt-jointed without joint filler. The weather being hot, the formwork was sprayed for 24 hours and more before the concrete was placed. The contractor was instructed on no account to treat or make good any faults discovered during or after the striking of the formwork.

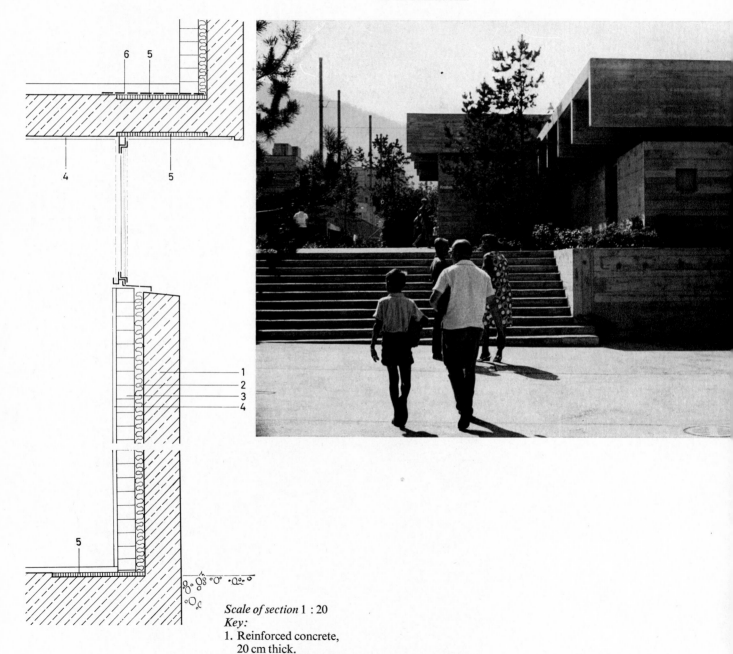

Scale of section 1 : 20
Key:
1. Reinforced concrete, 20 cm thick.
2. Glass-fibre matting, 5 cm.
3. Lime-sand bricks, 10 cm.
4. Plaster.
5. Compressed cork, 2 cm thick.
6. Damp-course.

The Recreation Centre at Heuried, showing the main entrance
looking towards the changing rooms

The Friis Residence at Braband, Denmark (1958)

Architects: Knud Friis and Elmar Moltke Nielsen, Aarhus
(*Photograph by Thomas Pedersen and Poul Pedersen, Aarhus*)

Upper-floor walls of 10 cm reinforced-concrete, untreated, backed by 8 cm mineral wool, aluminium-coated building-paper and wood panelling.

Grey Portland cement, sand 0 to 4 mm in diameter and gravel 8 to 12 mm, mixed in the proportions 1 : 2 : 3. Water/cement ratio 0·65. Crushing strength 240 kg per square centimetre (23·5 N/mm²).

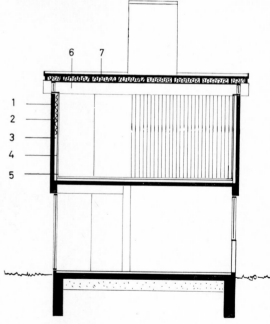

Scale of Section 1 : 1000
Key:
1. Reinforced concrete 10 cm thick.
2. Battening 4 × 6·5 cm.
3. Mineral wool 80 mm thick.
4. Building-paper.
5. Tongued and grooved match boarding, 2 × 12·5 cm.
6. Roof-joists 13 × 6·5 cm.
7. Mineral wool boards 10 cm, laid under building-paper and plaster boarding.

A House in Stuttgart–Bad Canstatt (1962–1963)

Architect: Werner Luz, Stuttgart–Bad Canstatt

(*Photograph by Gottfried Planck, Stuttgart*)

Outer walls 25 cm thick, in a single skin of lightweight concrete with Leca aggregate. The mix contained (per cubic metre) 270 kg Portland cement, 450 kg river sand 0–3 mm, and 300 kg of Leca 3–20 mm. Water/cement ratio 0·5.

No additional heat insulation was necessary. Nor was there need for a damp course since Leca absorbs practically no moisture.

Formwork of 10 cm-wide tongued boards, wrought both sides. No surface treatment. The inside walls also are of exposed concrete.

Points learned by experience included the following:
(1) Visual concrete made with Leca aggregate should be used only by thoroughly experienced contractors. It is always advisable to call in a consultant from the Leca suppliers.
(2) Wall thickness should not be less than 25 cm minimum.
(3) All reinforcement must be so disposed that particles of Leca do not adhere to it.
(4) The concrete should be laid plastically in layers of about 30 cm, and sealed.
(5) Since granules of Leca are lighter than water and therefore float, short lifts are essential.
(6) A sample of the concrete to be agreed before work begins.
(7) The source of supply of the river sand used should never be changed during the course of erection.

The "Park" Restaurant at Schaffhausen on the Rhine (1964–1965)

Architects: Walter M. Förderer and Hans Zwimpfer, Basle
(*Photograph by Brecht-Einzig, London*)

Outside walls of 30 cm-thick reinforced concrete, lined
internally with 5 cm of cork and finished with pinewood
panelling on 4 cm battens. Heat insulation by means of 5 cm of
cork arranged in 33 cm strips round the edges of all floors,
dropped into the formwork prior to placing.

Formwork of 12 cm-wide unwrought boards, chamfered
and grooved. No admixtures in the concrete.

The Wilhelm Lehmbruck Museum at Duisburg, in the Ruhr (1959–1964)

Architect: Dr.-Ing. Manfred Lehmbruck, Stuttgart

(*Photograph by Rüdiger Dichtel, Stuttgart*)

The formwork for the reinforced-concrete faces was of partly wrought, partly rough-sawn or sand-blasted boarding. On sections of the interior partitions, plywood shuttering was used.

The outer walls were constructed of two skins with a 2 cm-thick strip of expanded polystyrene laid between them, the skins being poured consecutively. Two coats of silicone fluate were applied as surface treatment.

A concrete wall some 8·40 metres wide by 10·6 metres high serves as optical background to the inner courtyard. The construction joints corresponding to the shuttering panels are staggered, in order to preserve the monolithic character of the building. The horizontal joints were not tightly stopped, and concrete leaking out was irregularly broken off when the formwork was struck.

Concrete mix: Light-coloured blast-furnace cement; aggregate up to 30 mm in size. Water/cement ratio 0·45. Fifty kilograms of quartz powder per square metre, and a water-proofing admixture in the proportion of 1 % of the cement content, were added to the mix.

Formwork of oiled hardboard. The tying-in of the reinforced concrete roof to the wall with its staggered construction joints presented some difficulty.

**The Lehmbruck Museum at Duisburg. The Sculpture Hall
and (*opposite*) an exterior view of one of the side walls
of the Museum**

(*Photograph by Zwietasch, Kornwestheim-bei-Stuttgart*)

The long walls of the Sculpture Hall reproduce boldly the
texture of the formwork—partly rough-sawn, partly sand-
blasted, and on purpose unevenly laid. The contrast between
this wall and the smooth curves of the exterior side walls
of the Exhibition Hall (*pictured opposite*) provides a good
illustration of the versatility of visual concrete.

The boards used in the Sculpture Hall were between 24 and
30 mm thick, of varying widths, and up to 7·26 metres (nearly
24 ft) long. The wall rises unbroken through two floors of the
Museum. It was constructed in two skins, with a 2 cm layer of
expanded polystyrene between them. The outside was treated
with silicone fluate.

The Exhibition Hall of the Lehmbruck Museum

is enclosed by two pairs of curving walls, each alternately concave and convex, facing one another across the Hall at either side as one comes in from the Entrance Hall. In order to increase the optical effect of the curvature, the boards used in the formwork were arranged to be of diminishing widths from the centre outwards.

Concrete mix as already described save that the size of aggregate was limited to a maximum of 15 mm, and that 50 kg more cement per cubic metre was used.

Formwork of planed boards, each rising to full shuttering height and varying in width from 22 to 2 cm, pre-treated with mould oil.

Scale of ground plan of Museum 1 : 1000
Key:
1. Entrance Hall.
2. Sculpture Hall.
3. Interior courtyard.
4. Sculpture court.
5 & 6. Space for visiting Exhibitions.
7. Picture Gallery.
8. Forecourt.
9. Lecture room (planned).

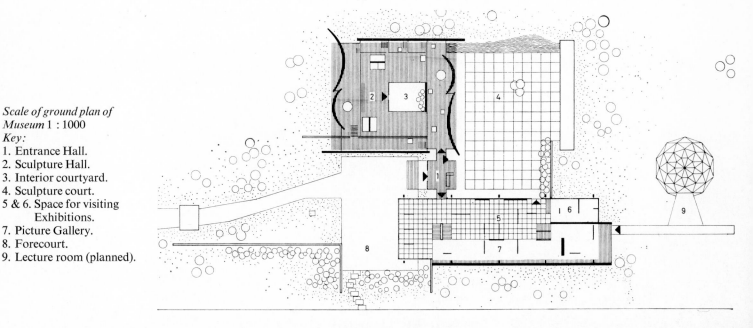

(*Photograph by Rüdiger Dichtel, Stuttgart*)

A Single-Family Residence in Westport, Connecticut, U.S.A. (1962)

Architect: John M. Johansen, New Canaan, Connecticut

(*Photograph by Robert Damora, Bedford Village, N.Y.*)

The Southern aspect of the residence

Scale of Ground Plan
1 : 300

1. Entrance Hall.
2. Living Room.
3. Recessed Fireplace.
4. Dining Room.
5. Kitchen.
6. Maid's Room.
7. Garage.
8. Dressing Room.
9. Bedroom.
10. Study.
11. Guest Rooms.
12. Kitchenette.

Exterior and interior walls are curved in the shape of an outer shell or crust. Emerging from them are the powerfully-ribbed profiles of the wall surfaces.

Concrete of light-yellow Portland cement; local aggregate with *Pozzolith* added.

Formwork of oak-veneered plywood. The sheets were cut by circular saw into strips of widths varying between 7·5 and 20 cm, butt-jointed, and deliberately left rough and unstopped. After striking of the formwork, faults were made good with a special mortar called "*Thorogeal*", and the whole was then coated with silicone. Careful making-good was essential in order to avoid penetration by frost, and to obtain an unbroken surface for application of the water-repellent paint.

Wall thickness 20 cm. Interior walls left partly as visual concrete, partly plastered, painted or papered.

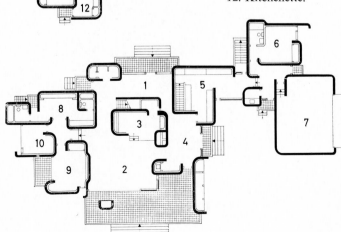

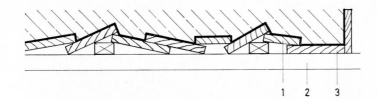

Details of the formwork
1. Vertical shutterboards.
2. Cross-battens in the formwork.
3. Right-angled boarding for door and window jambs.

The powerfully-articulated walls of the house at Westport rise from unribbed pedestals

(*Photograph by David Hirsch, New York*)

The living room of the residence in Westport, Connecticut
(*Photographs on this page and the page opposite by Robert Damora, Bedford Village, N.Y.*)

The fireplace recess is sunk into the floor of the living room
itself in the house at Westport

The Children's Zoo at Rapperswil, Switzerland (1961–1962)

Architect: Wolfgang Behles, Zürich

(*Photographs by Fritz Maurer, Zürich*)

The enclosure shown is divided into four paddocks for the animals, and is built of visual concrete throughout. Note how the exterior and interior shapes flow freely into one another in a way which is typical of the material.

The concrete mix contained a plasticizer ("*Plastocrete D600*"). The walls were tooled by pneumatic hammer, the roof left rough from the planed boards of the formwork.

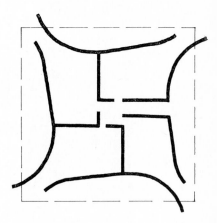

A free-standing wall of visual concrete at the Zoo has "thoroughfares" through it for big and small alike

A paddock in the Rapperswil Zoo

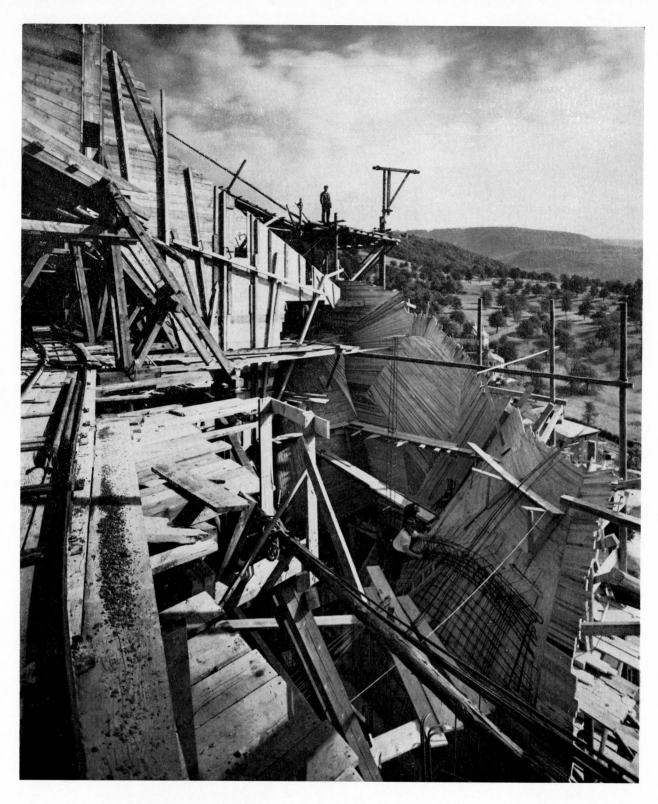

**The Goetheanum at Dornach, near Basle,
Switzerland (1925–1928). A view of the formwork on
September 30th, 1926**

Scale Model and Planning Advice: Dr. Rudolf Steiner
Structural Consultants: Leuprecht and Ebbell, Basle

(*Photograph by O. Rietmann-Haak, St. Gallen*)

Both the Planning Office and the Building Control Room
were situated at the Goetheanum itself, which was planned as
(and is still) the World Headquarters of the Anthroposophical
Movement founded by Dr. Steiner.

Two sections through the Dornach Goetheanum, on a scale of 1 : 100

The sections shown are reproduced from the formwork drawings done to a scale of 1 : 20 by the Basle engineering firm of Leuprecht & Ebbell. The shell of the outer wall is 10 cm thick in the sloping part of the roof and varies between 15 and 30 cm in the freely flowing walls.

On the original plans of May, 1927, the following notes occur:

(1) The figures in brackets correspond to the serial number in the bending schedules.

(2) The ribs of beams and hollow floors to be cast simultaneously with the slabs.

(3) Working joints to be agreed in advance.

(4) The cutting of all holes in finished reinforced concrete, together with any form of preliminary loading on them, to be agreed beforehand.

(5) Concrete mix: 250 kg cement per cubic metre of loose aggregate and sand of standard quality, in the proportions of 2 : 1.

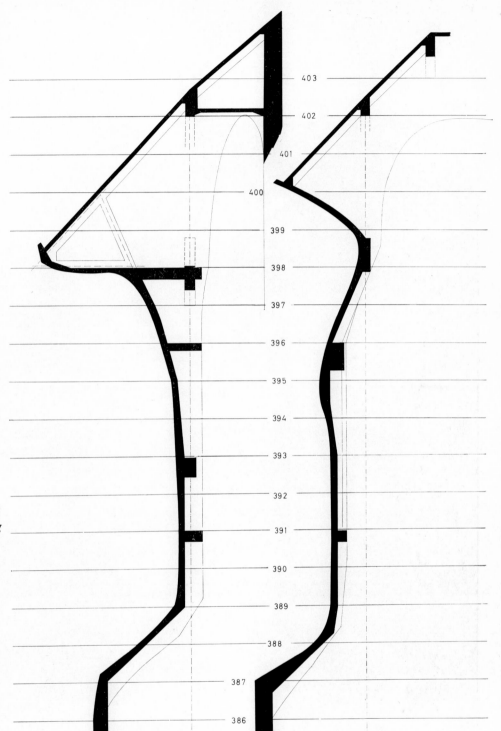

The Goetheanum was based on a model to the scale of 1 : 100 made early in 1924 by Dr. Rudolf Steiner himself. He had long been thinking explicitly of a concrete structure whose shape would be a development of the structural concepts on which the first Goetheanum of 1913–1923 had been based. It was from this model that the architects' office and the structural engineers developed the working drawings and the model of the finished building (which has an overall capacity of 110,000 cubic metres—nearly four million cubic feet).

The freely-sculptured walls are reinforced on the inside by ribs and tension members. The outer walls of the West Portico are in parts only 12–15 cm thick; but those outer walls which rise to a height of 30 m are up to 60 cm thick (though not everywhere reinforced).

On the inside the walls were originally merely rendered; but in 1964 those rooms in permanent use were insulated on the inside with 2 cm of expanded polystyrene or cork, with facing brickwork of hollow bricks 8 to 11 cm thick.

Dr. Rudolf Steiner's Goetheanum at Dornach
The West elevation as it appeared on September 29th, 1964
(*Photograph by Walter Grunder, Basle*)

The western stairway of the Goetheanum, showing the excellent condition of interior visual concrete after almost 40 years (*Photograph by Walter Grunder, Basle*)

In January, 1963, the then-83-year-old engineer, Ole Falk Ebbell, reminiscing about the building of the Goetheanum, recalled the day in 1927 when he showed Le Corbusier round the building.

"We had the plans," he went on, "and later the big model Mark 20, to which we had to adhere most strictly. Steiner was very exacting, but nobody interfered with us. We received the profiles for the exterior, and sometimes details of an internal roof-surface, but I had complete freedom as to what I did in between. . . .

"Reinforced concrete was at that time still in its infancy, and few people had experience of it. It had always been my principle to use as little concrete as possible. Never more than 16 litres (3·5 gallons) to the cubic metre, which is not much. But with water in those days we used to sin! Many a time we poured it in by the bucketful—50 or 60 litres at a time.

"But I wanted to learn from actual experience on the site, so I stood beside the man at the mixing machine and asked him questions. He had it all worked out. . . . After that, we made it much drier—but it must not be too dry. Forty litres is about right. It shouldn't be less—but it shouldn't be more either. . . . " (*From notes taken by the architect, Rex Raab.*)

The TWA Reception Building, Idlewild Airport, New York (1962)

Architects: Eero Saarinen and Associates, Hamden, Connecticut

(*Photographs by Ezra Stoller Associates, Rye, N.Y.*)

Four **Y**-shaped concrete supports carry the lightweight concrete roof shell, which is shaped to express the structural forces bearing on it. The area of the Reception Hall supported by these four supports is about 25,000 square metres (nearly 6½ acres).

**The passenger exit of the TWA Building at Idlewild provides a
close-up view of one the Y-shaped concrete roof supports**

In the thinnest sections of the roof shell of the building, the
thickness of the lightweight concrete of which it is constructed
is only 174 mm (less than 7 inches).

**The Architectural Faculty Block, Technical University
of the Middle East, Ankara, Turkey (1962–1965)**

Architects: Altug and Behruz Çinici, Ankara

Most of the buildings on the University campus were built of
exposed concrete. There was a special advantage here, in that
aggregate could be cheaply won on the site.

After a short training period, good visual concrete was
produced with unskilled labour.

Chapter 3

Visual concrete as an artistic medium

Condensed and adapted from an essay *Sichtbeton als Gestaltungsmittel* by Professor Max Bächer, holder of the *Lehrstuhl für Entwerfen und Raumgestaltung* at the Technical University of Darmstadt, Federal Republic of Germany.

All architects know that the physical properties of a building material greatly influence the way in which it can be processed and the purposes to which it can be put. In other words, every material possesses certain natural characteristics which are closely bound up with its inner structure. These characteristics limit its application, determine the tools and methods which can be applied to it, and affect both design and form. Custom and experience, by arousing certain feelings in the mind of the spectator, cause one material to be considered suitable for certain tasks, but not for others. New tools and working methods, together with changes in cultural or economic conditions, often open up new fields of application for a given material as its hitherto-neglected properties begin to achieve recognition.

The architect is therefore compelled to be always openminded, and should never reject new building materials or new building techniques merely because he himself has so far had no direct experience with them.

The relationship of form, content and material

There does not exist a building material which is *intrinsically* either ugly or inferior; for there is none which cannot achieve beauty in the hands of a designer using it in the right conditions and in the right place. Conversely, even the most valuable of materials can produce a lifeless, even a repellent, effect when they are wrongly used or wrongly applied.

Material, form and content represent always a living, organic and indivisible whole. Only in imagination can one separate them from one another, and so perceive the part which each plays in the whole or why in one case a work gives an impression of completeness yet in another displays a lack of balance. For this reason, the technological characteristics of a material can never be regarded as the sole criterion by which its suitability for use can be judged. It is just as important to know whether that material is capable of expressing a certain content in relation to the creative imagination. A convincing result is only achieved when all the factors combine to present an intellectual, formal and material unity.

Form and content, in particular, cannot exist by themselves alone. The content is unavoidably linked with a specific form; and there is no form which does not contain or conceal content. Both, however, need a material in which to express themselves. Form and content are contributed by the artist, but the material itself is not. No sculptor or architect can create his own material. He can only use that which comes to him in the pre-determined shapes of industrial products—moulded brick, sawn beam, cut stone or extruded section—which he can then select, remodel and manipulate.

Concrete, on the other hand, by its very nature lacks inherent form. Placed in the liquid state, it takes the shape of the mould in which the mixture has hardened. The material thus emerges directly from its ingredients in its finally valid form. Creation of material and creation of form take place simultaneously, and the original shape of the material must also be the final form of the building.

Herein lies the essential contrast between visual concrete and the more traditional prefabricated building materials. Whereas the architect cannot influence the material form of the industrial product (or can at best only do so indirectly), concrete undergoes at his hands its very creation. He can exert his influence over his material at the very moment when it is coming into being; and he can adjust its properties and characteristics very closely to the requirements of his design.

A unique building process

Since it can assume the shape of any mould, the plasticity of concrete is limited only by the technical characteristics of the formwork in which it is placed. Maximum freedom in choice of shape is thus linked with the stern discipline of virtual unalterability once that choice has been made.

The crucial point in construction, normally the erection of the building itself, is in the case of concrete the erection of the mould. The operation of placing is not in itself important, for it serves merely to fix permanently the plastic outline of a temporary concavity.

The uniqueness of this process calls for the greatest possible care in the planning and construction of the formwork, for once that is finalized, all possibility of alteration or improvement to the emerging shape is excluded. Herein lies the essential difference between building in visual concrete and building with the traditional construction materials.

There emerges from this an indication of clear duality in the nature of a building in visual concrete. Unlike the usual hand-made product which receives its final form only when the last touches are put to it, the operation of concreting demands that every factor affecting ultimate shape be fixed beforehand and every detail planned from the start.

Here, then, is a characteristic typical of a mass-produced unit of output from a long production-line built inextricably into a structure which is by its very nature an unrepeatable prototype! And a prototype, moreover, which by virtue of its size, its cost and its relative indestructibility, can hardly ever be "thrown out as a reject" if it fails to come up to standard.

A task for the skilled craftsman

Concrete, of course, is widely used in the production of standardized building components. In this form, it is of little interest from the point of view of its value as an artistic medium. Designed, as they must be, for multi-purpose applicability, pre-fabricated components in any material are bound to take on the featureless anonymity of the mass-produced article.

But how different is the erection of an entire building from concrete placed on the site itself! Here is the very essence of a "hand-made" article. Here is a task calling, if ever one did, for the fullest mental and manual skills of the dedicated craftsman.

Never before have the architect designing a building and the craftsman carrying out that architect's ideas been called upon to plan their work and to bring it into existence by first creating a *negative mould* of what they want to achieve, a mould complete and accurate in every detail and every dimension—and a mould which must itself be destroyed before the positive form of the structure itself can become visible for the first time!

The demand for creative freedom

Critics, perhaps, will say that such a call on the highest skills of craftsmanship is an anachronism in an age of industrial mass-production and disappearing handicraft skills. Can the growing interest which a single generation of architects is showing in so labour-intensive a material as visual concrete possibly be justified?

The answer surely lies in the modern artist's insistent demand for creative freedom, for release from the constrictions imposed on him by traditional methods and intractable materials. This demand is being expressed in many other fields of Art today. Is it altogether fanciful to offer an explanation of why this is so?

Our grandfathers, who felt themselves by comparison with ourselves to be free men, had sober taste in their pictures, in their sculpture and in their standards of artistic appreciation generally. The artists of today, oppressed (it may be) by the dread of anonymity in face of the outpouring of soul-less perfection from the machine, feel compelled to experiment ever more boldly in their choice of materials and in the uses to which these materials are put. Freedom to create, freedom to experiment—that is the demand—that is perhaps the psychological need of our harassed generation. . . .

How fortunate, if this be so, is the architect in having at hand a material so infinitely flexible, so literally limitless in its plasticity, as is visual concrete. . . .

An untreated surface essential

Perhaps the most essential requirement of a building in exposed concrete is that its surface after casting should be left entirely untreated. When the shuttering is removed, the concrete surface will faithfully reflect every knot and every line of the grain of the timber against which it has hardened, and every crevice between individual planks into which the concrete oozed when it was still in its liquid state.

Any attempt to "improve" this surface offends against the very nature of visual concrete. It may be true that processes like washing, or sandblasting, or bush-hammering, sometimes reveal additional possibilities in the material itself, such as variations in its granular structure or in its colour. Yet it remains true that any surface treatment of exposed concrete after it has been placed not only destroys the very features which are most typical of the material itself, but also shows up with relentless clarity any mistakes there may have been in the casting. In this context the rule that materials must be allowed to remain absolutely true to themselves and to their own nature is (despite its moralistic ring) appropriate.

Especially to be deplored, in the writer's opinion, is any attempt to "improve" the cast surface by applying paint or by rendering it with any form of plaster. Analogies drawn from Doric temples of the seventh and sixth centuries B.C., where the natural stone was sometimes covered with a fine layer of stucco, which was then probably painted, are not appropriate to the more solid bulk and monolithic structure of modern buildings in concrete.

Ability to age gracefully

One of the charms of exposed concrete (and another excellent reason why it should never be rendered or painted) is that it tends to improve with age. As the years pass, the surface of a structure in visual concrete acquires a living patina which gives it added liveliness and expression.

This ability to age gracefully makes concrete a material especially suitable for the building of structures intended to be lasting. In this, it is sharply to be distinguished from the smooth, expressionless surfaces of other materials whose technical nature precludes the possibility of visible ageing. Though the purpose behind such façades is that they shall always look modern and up-to-date however long they endure, yet their very shapes will always unerringly betray the date of their construction.

A solid cast structure is quite different. There it stands, and will always stand—final, unrepeatable, timeless in its immutability, and ageing with grace. Indeed, had it not been for a factor which we must now briefly examine, concrete would have been the perfect building material for every despot down the ages!

Prejudice against concrete as a building material

Although concrete was known to the Romans under its Latin name of *caementum*, the technique required to make it usable for structural purposes was not available until the invention of reinforced concrete. Thereafter, its admirable structural properties caused it for a long time to be left only to the structural engineers. These latter appreciated its advantages, and made allowances for the peculiarities of its technical properties.

To architects, however, the use of concrete in their work was for a long time suspect. It lacked any form of its own; and its difficult technical and structural peculiarities were so obvious that its use tended to be restricted to outlets in which the architect's part was all but negligible.

The word "concrete" came to be equated with the poor and the sordid. The new material fitted badly into the romantic world of the summer-house and the conservatory—and the despot has yet to rule who wished the monuments he builds as lasting memorials to his own vainglory to be considered by his subjects as poor or sordid!

In Germany, it may be worth adding, there was allied a further form of prejudice against concrete which can only be called "ideological". To the Third Reich, concrete was objectionable because it lacked any specifically "Germanic" origin as a structural material. This rejection as an "international" material persists to some extent even today; and one cannot help suspecting that behind some of the objections which are put forward to the technical and architectural properties of concrete (insufficiently explored though these properties still are), there often lurk prejudices of an ultimately socio-political nature.

Affinities with cut stone

Despite the entirely different processes by which the two materials are worked, concrete possesses certain properties which resemble those of stone itself. Both materials possess weight and solidity. Both are strongly resistant to external influence, yet both are capable of being altered to a desired shape. Both convey the impression of permanence and indestructibility, and as such are fit material for architecture which is intended to be durable and for the construction of buildings intended to give monumental expression to the convictions of a social age.

Even in the effects they produce when viewed in unfinished form, structures in stone and in visual concrete possess marked affinity. When the pyramids, the temples and the cathedrals of old were built, their architectural shapes became clearly defined as soon as the rough masonry was erected. Subsequent embellishment did little to alter their basic structural form; it merely exploited the possibilities latent in it. With concrete also (by contrast with those hybrid "assembled" buildings in which the basic architectonic form often does not emerge until the final stages of construction), the creative idea again becomes apparent in the very structural substance of the "unfinished" building.

A danger to guard against

There remains to be considered one danger that is liable to arise in the use of visual concrete, and one type of building for which its use is functionally unsuitable.

The danger is this. Just as technically skilled stonemasons could abuse the natural properties of stone by over-elaboration in carving it (witness, for example, those filigree sections of Gothic church windows which were sometimes made so thin that they needed iron supports to preserve them from collapse), so can the use of structural concrete be over-strained to the point of virtuosity. There is sometimes a failure to acknowledge that a material is being forced to assume a form basically foreign to itself. The more difficult the technical task which the virtuoso sets himself, the more welcome he often finds it. In this his aim (conscious or unconscious) is to display the faultlessness of his own personal skill, and to show off the command which a technically brilliant craftsman can exercise over the material with which he is dealing.

This is a danger to which the worker in visual concrete is particularly exposed by reason of the very freedom of choice of form which his material allows him. It is therefore of the first importance that the means employed in shaping the concrete must always be strictly related to the nature of the task set.

An unsuitable use for visual concrete

One use for which visual concrete is seldom suitable is in the construction of municipal buildings. Perhaps the principal demand made on the municipal planner of today is that the buildings he puts up shall be flexible enough to be altered in accordance with rapid and unpredictable changes in the needs of the society which the buildings are intended to serve. Since the "life expectancy" of such buildings (that is to say, the period for which they are likely to be suitable for the use for which they were erected in the first place) is liable to be comparatively short, it has to be accepted that freedom of future planning must always be an important element in their design.

It must equally be accepted that buildings made of concrete cast *in situ* seldom conform to this paramount demand. In such buildings it is almost impossible to marry the creative urge of the artist with the practical requirements of the planner.

It is indeed unfortunate for the municipal architect that this need for maximum adaptability in so many of the buildings he designs should have arisen at a time when his long-suppressed urge for plastic expression is causing him to turn to one of the most inherently permanent of all materials.

Enthusiasm—but tempered with scepticism

In conclusion, one would say that building in visual concrete raises problems which are intimately connected with the importance of architecture to this age of ours. The unrepeatable singularity of the casting itself, the ability of the concrete to age gracefully, its permanence, the craftsmanlike skill needed to produce it, the multiplicity of the forms which it is able to take—all these must ensure for visual concrete an important place among the media of creative art in the modern world.

To use it successfully, however, demands a clear intellectual appreciation of its limitations. In its very plasticity lie both its attractions and its dangers; and the architect's legitimate enthusiasm for a method of construction which has opened up for him new possibilities of creative thought within new dimensions needs to be constantly moderated by practical scepticism about the tasks which this new medium is capable of performing with success.

Chapter 4

Framed structures in visual concrete

OPINIONS

For—

"Reinforced concrete is the finest building material man has yet invented. The facts that it can be moulded into practically any shape, and that it stands up to any strain, border on the miraculous. In the realm of building, it has removed all limits to the creative imagination." (*Pier Luigi Nervi*)

—And Against

"Concrete as an artistic medium makes sense neither formally, nor structurally, nor economically, nor physically. Concrete is a conglomerate which can be cast into any mould, made—to put it plainly—of muck in which the sometimes attractive and naturally-formed aggregates are polluted by being dropped into a dank, synthetic sauce." (*Raimund Probst*)

**Housing Unit at Marseilles (1948–1952) :
The Eastern Elevation**

Architect: Le Corbusier

(*Photograph by Lucien Hervé, Paris*)

The concrete is of standard mix, left natural and rough from the shuttering. Precise drawings provided for the formwork showed how the layout of even individual shuttering-boards followed the architect's celebrated "Modulor" system. Since the whole building is based on this system, the panels could be used over and over again without alteration.

The Housing Unit at Marseilles, by Le Corbusier
(*Photograph by W. Faigle, Stuttgart*)

An example of how "the Master" managed the transition from
functional form to free sculpture.

**A close-up view of Le Corbusier's
"Unité d'habitation" at Marseilles**

(*Photograph by Lucien Hervé, Paris*)

Unwrought boards 18–22 cm wide were used for the formwork.
They were butt-jointed at random within the chequer pattern.

The Capitol at Chandigarh in the Punjab (1951–1963)

Architect: Le Corbusier

(*Photographs by Lucien Hervé, Paris*)

The Secretariat Building houses both Ministries and the Regional Administration

The City Centre of the newly-created capital of the Punjab contains, amongst other buildings, Le Corbusier's Secretariat (in which the Ministries are housed), the Parliament Building and the Law Courts.

All these buildings were designed in rough exposed concrete made with sheet-metal shuttering. The dimensions were based on the "Modulor" system throughout. All curving walls were cast in wooden formwork.

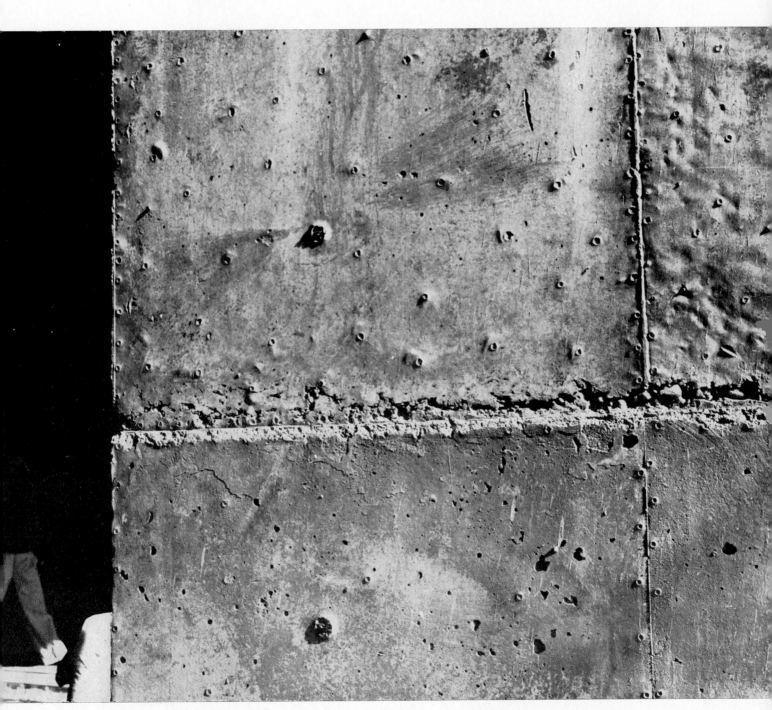

The face of the concrete reproduces the very nail-holes in the sheet-metal shuttering. The positions of tie-wires are also visible.

The Law Courts in Le Corbusier's Chandigarh
Note how, as a measure of protection against the heat, the roof is raised to allow the breeze to pass freely under it.

Le Corbusier's Chandigarh

The Parliament Building at Chandigarh

An Interior in the Parliament Building at Chandigarh
Note how the homogeneous nature of the concrete used as
building material throughout assists the informal merging of
interior and exterior space.

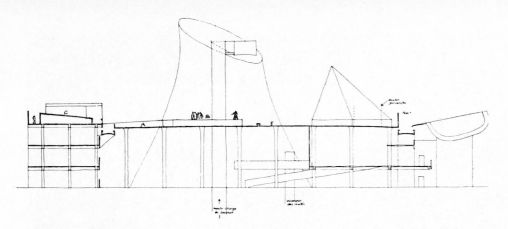

This section drawing of the Parliament Building is taken from Le Corbusier's *Œuvre complète 1952–1957*, published in 1957 by Editions Girsberger, Zürich.

The Parliament Building of Le Corbusier's Chandigarh : Interior View

(*Photograph by Lucien Hervé, Paris*)

The core of the Parliament Building is a vast circular Assembly Hall, dominated by a soaring light-shaft. A colonnaded outer hall, three storeys high supported on slender columns, surrounds the Assembly Hall itself.

**The Church of Notre Dame at Le Raincy, France
(1922–1923)**

Architect: Auguste Perret

(*Photograph by Chevojon, Paris*)

Note the contrast between the starkness of the concrete
columns in Perret's famous Church above, the elegance of Le
Corbusier's work at Chandigarh on the opposite page, and
the vigour and technical skill of Michelucci's modern work in
the Tuscan Church overleaf.

**The Church of St. John the Baptist, off the motorway
near Florence (1963–1964)**

Architect: Giovanni Michelucci, Florence.

The roof supports, spreading their limbs like trees in concrete,
call to mind the architecture of Gaudi.

The Pirelli skyscraper, Milan (1960–1961)

Architects: Professor Gio Ponti, Antonio Fornaroli and
      Alberto Rosselli, Giuseppe Valtolina and Egidio
      dell'Orto, Milan
Structural Engineers: Pier Luigi Nervi (Rome) and Arturo
      Danusso, (Milan)

(Photograph from "Domus" No. 379, Milan, 1961)

Situated on top of 30 floors of the concrete skyscraper is a two-storey service area whose purpose is to house the technical installations of the building.

Clearly shown in the photograph are the strongly articulated structural elements in visual concrete so characteristic of Nervi's work.

The Ponthieu Garage in Paris (1906)

Architect: Auguste Perret

(*Photograph by Chevojon, Paris*)

Perret's garage, a building contemporary with the earliest
work of Frank Lloyd Wright in America (*see page 17*), is a
landmark in the architectural use of concrete almost as
celebrated as his church at Le Raincy, illustrated on page 75.

    The building forms a striking contrast to the monumental
grandeur of Paul Rudolph's dramatic "Autotemple" pictured
on the opposite page.

A Multi-storey garage for 1500 cars at New Haven, Connecticut (1963)
Architect: Paul Rudolph, New Haven
(*Photograph by Ezra Stoller, New York*)

Pomp and permanence are graven on the very concrete of this worthy Cathedral for the chrome and metal Gods of Today!

**The Monastery of La Tourette,
near Lyons (1956–1960)**

Architect: Le Corbusier

(*Photograph by Robert Winkler, Stuttgart*)

The photograph below (*by Lucien Hervé, Paris*) shows
a view into the inner courtyard of the Monastery. To the
right is one of the conversation cells, strongly sculpted in
sprayed concrete. The prism-shaped objects in the centre of the
picture are openings letting light into one of the side chapels of
the Church, whose main wall is seen in the background.

As was often Le Corbusier's practice, strongly contrasting
forms and surfaces are set abruptly side by side. Each of them
has its own significance in the composition as a whole, but has
little meaning by itself.

The Southern Elevation of the Monastery at La Tourette

(*Photograph by Robert Winkler, Stuttgart*)

The three levels of the Monastery seen in the photograph above comprise, respectively, a foundation and access zone open to the air; the refectory; and monks' cells.

Each level is as sharply differentiated in form as is the use to which the concrete is put in shaping it. *Below:* Heavy slabbed pillars. *Above:* Slender ribs of concrete hold in place the unframed glass of the refectory windows.

The cantilevered balconies of the upper floors have pre-cast balustrades faced with rough-cast brushed-concrete slabs.

Concrete B 300 of grey Portland cement with plasticizer admixture. Formwork of unwrought boards of random widths, with all joints left unstopped to enhance the effect of rawness.

The walls of the penthouse are of 12·5 cm thick reinforced concrete, heat-insulated on the inside with 7·5 cm of rendered woodwool. The doors and windows are painted cobalt blue.

Sound insulation between the factory and the offices and sales areas was satisfactorily achieved by means of a 20 cm thick reinforced concrete wall. All non-load-bearing partitions are of removable prefabricated concrete.

**The factory at Thun: part elevation showing the living and
sleeping quarters of the penthouse**
(*Photograph by Leonardo Bezzola, Berne*)

The private house of the architect R. G. Otto, in Liestal, Switzerland (1960)

Architects: Förderer, Otto and Zwimpfer, Basle

(*Photograph by Bärtsch, Liestal*)

The large living-room, with its gallery serving as a study and its fireplace open on three sides, forms the pivot of everyday life.

The reinforced concrete of which the house is built is exposed even on the inside walls. The windows, partitions and panelling are of pine; the floor of artificial stone tiles with strip carpeting.

A view of the Town Hall at Asker, Norway (1961–1963)

Architects: Kjell Lund and Nils Slaatto, Oslo

(*Photograph by Bjoern Winsnes A.S., Ullern*)

The entire building was cast *in situ*, the dark colour of the concrete being achieved by using nearly black limestone aggregate, with the addition of iron oxide.

The concrete mix, per cubic metre, was as follows: Standard Portland cement, 375 kg; sand, 710 kg; crushed gravel of 13–20 mm diameter, 1065 kg; iron oxide, 21·5 kg; water, about 175 litres with a plasticizer added.

Exterior surfaces were point-tooled by hand; interior walls either similarly point-tooled or bush-hammered.

The three sectional drawings on the next page show details of the balustrading attached to (*on the left*) the standard flooring of the main building and (*on the right*) the flooring of the single-storey annexes.

Key to Section Drawings on page 86.
Scale 1 : 20
  1. Bush-hammered concrete.
  2. Scraped concrete.
  3. Concrete left rough from the formwork.
  4. Outer shell of 20 cm concrete.
  5. Sheet of corrugated asbestos cement.
  6. Ventilation.
  7. Lightweight concrete, 15 cm thick.
  8. Inner shell of 10 cm concrete.
  9. Floor covering laid on 5 cm screed.
10. Foamed plastic (Isopor) 2 cm thick.
11. Lightweight concrete, 7 5 cm.
12. Cantilevered concrete floor.
13. Air space.

Detail of the façade of the Town Hall at Asker

A point of interest is the rebated transition between the flat surface of the lower edge of the flooring and the heavily-tooled balustrade.

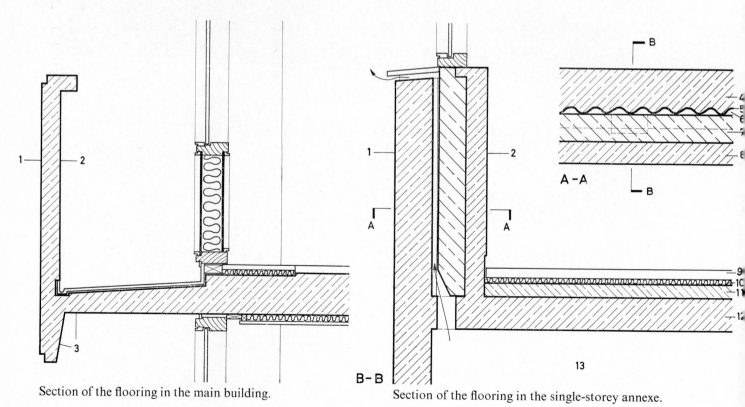

Section of the flooring in the main building.

Section of the flooring in the single-storey annexe.

The Council Chamber in the Town Hall at Asker

The walls are of dark-coloured concrete, heavily tooled. Here, as in other Council chambers, the dignity of the room is enhanced by the quality of the hand-worked craftsmanship.

Photograph by F. Murasawa, Shokokusha, Tokyo

Left: **Detail of a corner of the front wall**
(*Photograph by T. Taira, Tokyo*)

The Town Hall of Kurashiki, Japan (1958–1960)

Architect: Kenzo Tange, Tokyo

The Town Hall is a framed building having a ground-plan grid of
1·80 m. The wall in which the windows are set is screened by a pattern of
prefabricated concrete elements which not only protect the windows
from glare but impart a vivid articulation to the façade as a whole.

The vertical dimensions of these prefabricated elements were varied
between 245 mm, 410 mm, 675 mm and 1105 mm (about 10″, 16″, 26″
and 42″) so as to allow a number of different combinations (*see
illustration, left*). On blank wall surfaces the elements are some 10 cm
thick. They become trough-shaped, with depths varying between 410,
575 and 1105 mm, in areas of heavy surface indentation.

The principal staircase in the entrance hall of the Kurashiki Town Hall
(*Photograph by Retoria/Y. Futagawa, Tokyo*)

The City Hall at Hiraoka, Japan (1963–1964)

Architect: Junzo Sakakura, Tokyo

(*Photographs on both pages by Retoria/M. Outsuka, Tokyo*)

Concrete mix: 306 kg light-grey Portland cement, 761 kg sand, 1081 kg aggregate, 178 kg water per cubic metre (say, 517, 1280, 1830 and 300 lb per cubic yard, respectively).

  Formwork of wrought cedar-wood boards about 10 cm wide, tongued and grooved. Visual concrete surfaces either silicone-painted or bush-hammered. On internal walls, 2 cm plaster. No special heat insulation because of the warm climate.

Scale of Section (*right*) 1 : 100

The main entrance to the
City Hall, Hiraoka

On the framework of the building the
concrete was left exactly as it emerged from
the formwork. The walls themselves were
bush-hammered

The Metropolitan Festival Hall, Tokyo (1961)
Architects: Kunio Mayekawa & Associates, Tokyo
(*Photographs on both pages by Yoshio Watanabe, Tokyo*)

The concrete mix per cubic metre was 327 kg Portland cement, 692 kg sand and 1139 kg aggregate. Formwork of wrought boards. No surface treatment after it was struck.

Note how the lines of junction between the large areas of concrete cast in alternate vertical and horizontal shuttering are marked by vertical expansion joints and strongly accented horizontal grooves.

The Metropolitan Festival Hall, Tokyo (1961)
Architects: Kunio Mayekawa & Associates, Tokyo
(*Photographs on both pages by Yoshio Watanabe, Tokyo*)

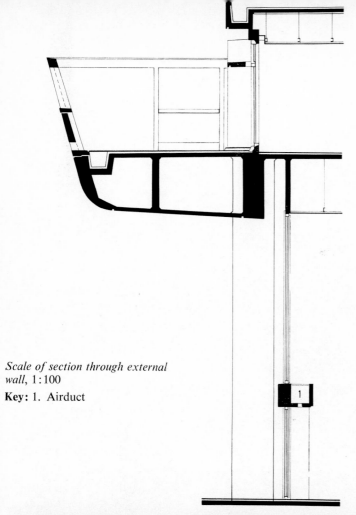

Scale of section through external wall, 1:100

Key: 1. Airduct

The Concert Hall of the Tokyo Metropolitan Festival Hall

Note the three-dimensional decorative patterning which breaks up the surface of the side walls, and at the same time improves the acoustics of the Hall. Strongly accentuated by side-lighting, this patterning contributes to the festive atmosphere of the interior.

Theatre Festival Hall at Ingolstadt, Bavaria (1963–1965)

Architects: Hardt-Walther Hämer and Marie-Brigitte Hämer-
Buro, Ingolstadt
Associates: Klaus Meyer-Rogge, Norbert Weber, Hans
Schubert
(Photographs on the two facing pages by H. W. Hämer, Ingolstadt)

The entrance elevation

The foyer opens on to the terrace through 7 metre-high (23 foot) suspended plate-glass panes.

'The striking of the formwork which moulded the large expanses of visual concrete needed to be done with the greatest care,' remarked the foreman of the construction gang.

'It must be strongly impressed on the men that efficient recovery of the shuttering material is *not* the principal object of the operation.'

94

In the Festival Hall at Ingolstadt a stairway in visual concrete leads up from the Exhibition Halls to the Terrace

Diagrams illustrating the sequence of operations at the end of one lift and the beginning of the next

A. At well-defined horizontal joints.
1. Complete the first lift of concrete. Insert a recessed dovetailed batten (**a**) of dimensions $3 \times 2 \times 1 \cdot 5$ cm.
2. Strike the formwork of Lift I, leaving the batten in position.
3. Wire the formwork for Lift II firmly to the ends of the formwork of Lift I which have been left in position for the purpose. Insert strips of plastic foam (**b**) behind the batten as a protection against oozing grout.
4. Pour Lift II.

B. At invisible horizontal joints.
1. Pour Lift I, inserting on top of it battens (**c**) 3×5 (or, better still, 4×6) cm in diameter.
2. Strike the formwork, leaving only the boards required to tie in Lift II.
3. Wire the formwork for Lift II firmly to the ends of this shuttering, and insert the plastic foam strips (**b**). The joints between the boards of the new formwork must be precisely calculated so that the rhythm of the plank breadths reproduced on the finished concrete of Lift II carries on unbroken from the corresponding impressions on the face of Lift I.

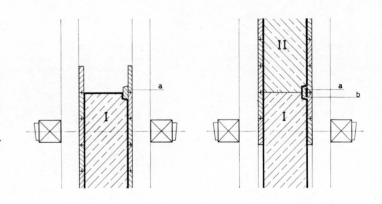

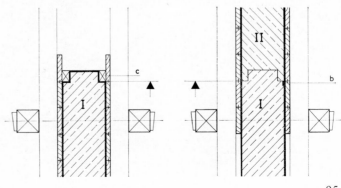

The Concert Hall in the Theatre at Ingolstadt

(Photograph by Wagner, Ingolstadt)

The concrete used in the construction of the Festival Hall was made of grey-green cement, 320 kg to the cubic metre, mixed with four sizes of aggregate, 0 to 3 mm, 3 to 7, 7 to 15 and 15 to 30 mm, and with 420 to 450 kg of powdered stone per cubic metre. The water/cement ratio varied between 0·4 and 0·5.

The formwork was of 24 mm-thick pine boards 14 cm wide, machine-planed, tongued and grooved, and treated with mould oil.

The outer walls were of 25 cm-thick reinforced concrete lined with 3·5 cm of Isotex, 3 cm battening and 1 cm plasterboard with an air-space behind. In the kitchens 5·1 cm of foam glass was used instead of the Isotex to serve both as insulation and as damp course.

Expansion joints were inserted every six to nine metres on the exposed concrete surfaces. The latter were given a coating of colourless matt silicone wash as soon as the formwork was struck.

Note the adjustable panels suspended under the ceiling of the Hall to assist in acoustic control

The photograph to the right (by *H. W. Hämer, Ingolstadt*) shows the foyer of the Theatre, with its pendant clusters of 15-watt lamps.

The auditorium and stage of the Theatre in the Festsaal at Ingolstadt
(*Photograph by the Donau Kurier, Ingolstadt*)
The indirect lighting of the walls illuminates murals in gold leaf by Heinrich Eichmann, of Zürich. The colour chosen for the seat-covers in the stalls was terra-cotta.

The St. Gallen (Switzerland) Technical College for Economics and the Social Sciences (1960–1963)

Architects: Walter Förderer, Rolf G. Otto and Hans Zwimpfer, Basle and St. Gallen

(Photographs on this and the facing page by Fritz Maurer, Zürich)

All the fixed structural elements of the College buildings are finished in visual concrete, both inside and out. Movable panels on outer walls are of painted steel, while all interior partitioning is constructed of painted timber.

Note how well the cantilevered staircase pictured below reveals the structural and formal possibilities of reinforced concrete. All those details of construction which are normally left to be dealt with when the main structural work has been completed must be thought through at a much earlier stage; for it is not possible to chase wiring conduits and the like into exposed concrete walls and to conceal the slots with plaster. All tubing of this kind must be cast into walls and ceilings at their very 'birth'; and every detail of the complete architectural concept must be built into the basic structure of the edifice.

The Inner Courtyard of the Technical College at St. Gallen

The sculpture by Hans Arp in the foreground, together with
the cube-shaped concrete forms dotted about the
Courtyard itself, establish a connecting link between the
otherwise-unconnected College buildings seen in the picture
below.

A commercial and residential building at Emsdetten, North-Rhine Westphalia, West Germany (1965)

Architects: M. C. von Hausen and Ortwin Rave, Münster

(*Photographs on this and the facing page by Hans Eick, Emsdetten*)

Concrete B225 of blast-furnace cement and gravel. Formwork of unwrought boards, 12 to 14 cm wide, which had already been used once before. All horizontal surfaces were protected against moisture by a coating of transparent polyester resin. Heat insulation by means of foamed plastic covered on one side with woodwool boards.Two coats of rendering.

The architects believe that their solution of the problem of wall-insulation fell short of the ideal. "The use of foam-glass, which we originally proposed, was ruled out on grounds of expense. There is no insulating material on the market which incorporates a damp course and is at the same time suitable as a backing for plasterwork."

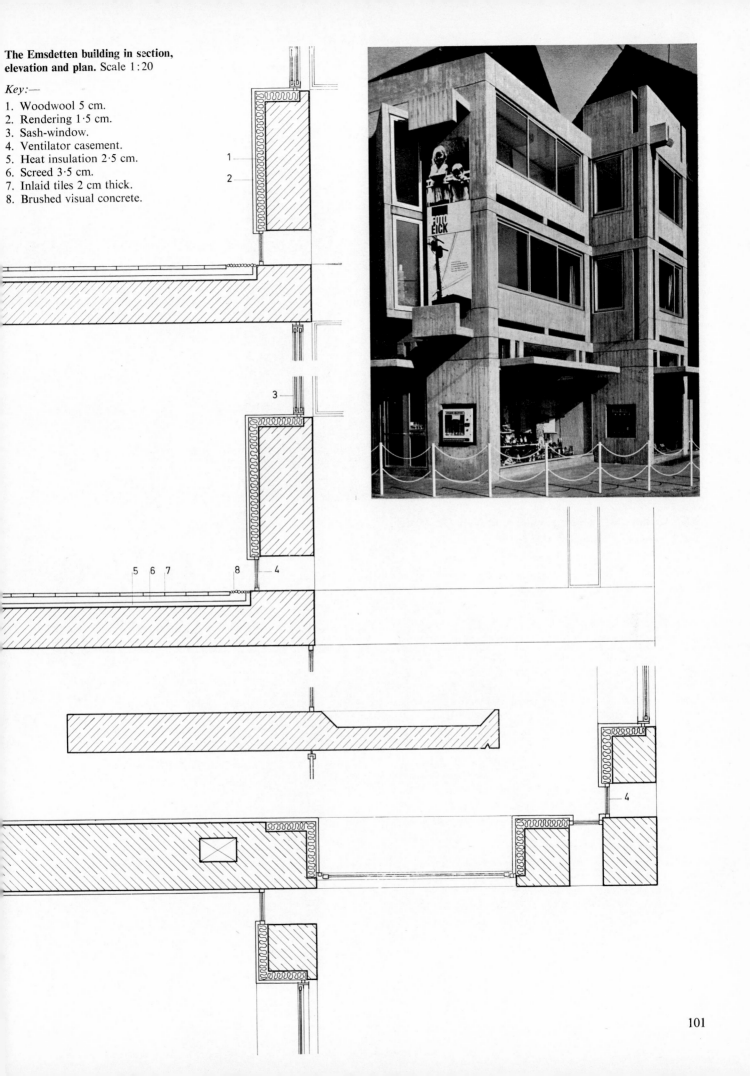

The Emsdetten building in section, elevation and plan. Scale 1:20

Key:—

1. Woodwool 5 cm.
2. Rendering 1·5 cm.
3. Sash-window.
4. Ventilator casement.
5. Heat insulation 2·5 cm.
6. Screed 3·5 cm.
7. Inlaid tiles 2 cm thick.
8. Brushed visual concrete.

The dimensions of the building are 49 × 38 metres. Maximum height is 28 metres.

The Art and Architecture Building at Yale University, New Haven, Connecticut (1961–1963)

Architect: Paul Rudolph, New Haven

(Photographs by Ezra Stoller Associates, Rye, N.Y.)

Concrete made of light-coloured Portland cement, with sand and gravel up to 4 cm in diameter. To ensure a uniform mix for the entire building, the sand and aggregate for the entire job were ordered at the same time, and stored. The formwork for the walls consisted of 60 cm-wide blockboard panels with trapezoid battens nailed to them. The formwork was oiled, and struck not later than the day after casting.

Exterior walls are 30 cm (12 in) thick, with no heat insulation or damp course. The door fixtures were inserted into the concrete without frames, and the sash windows were secured at top and bottom by aluminium sections let into the concrete. The pipes for all the plumbing run in hollow-cast columns.

The ribbed surface produced by the formwork was worked over after striking, the proud faces being knocked off by hand-held hammer as soon as the concrete had hardened sufficiently to allow the pebbles to be split without being broken out. To achieve the most uniform effect possible, every second concrete ridge was worked over from the top downwards, with the intervening ridges being tackled from the bottom upwards.

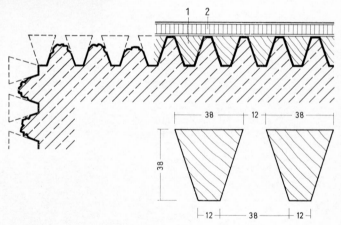

Detail of the shuttering:—Scale 1 : 5

Key: 1. Wooden battens. 2. Plywood board.

The profile of the wooden battens is shown enlarged to a scale of 1 : 2.

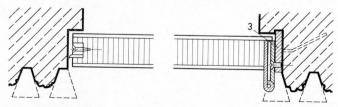

Detail of door jamb:—Scale 1 : 5

Key: Steel flat, 55 × 10 mm in section and 160 mm long, embedded in the concrete.

A close-up of the hand-hammered ribbed-concrete surface of the
Art and Architecture Building at Yale

The Old People's Home and Ambulance Station of the German Red Cross in Karlsruhe, West Germany (1965–1966)
Architect: Dr.-Ing. Reinhard Gieselmann, Karlsruhe
Associate: Peter Skowronek.

(Photographs by Thilo Mechau, Karlsruhe, taken before building was completed)

Several different concrete mixes were used to comply with varying structural requirements. Standard Portland cement was combined with aggregate graded in the following proportions: 0–3 mm, 35%; 3–7 mm, 18%; 7–15 mm, 23%; 15–30 mm, 24%; plus 40 kg of 0·06–0·2 mm quartz sand per cubic metre to compensate for eventual changes in the colour of the cement.

The waxed formwork was of 9 cm wide tongued-and-grooved boards, planed for the main elevation, rough-sawn for the elements standing 30 and 60 cm proud from it. No surface treatment was applied to the finished concrete.

The close-up view to the right shows how the façade of a building can be given exciting plastic form even when building regulations allow little scope for variety

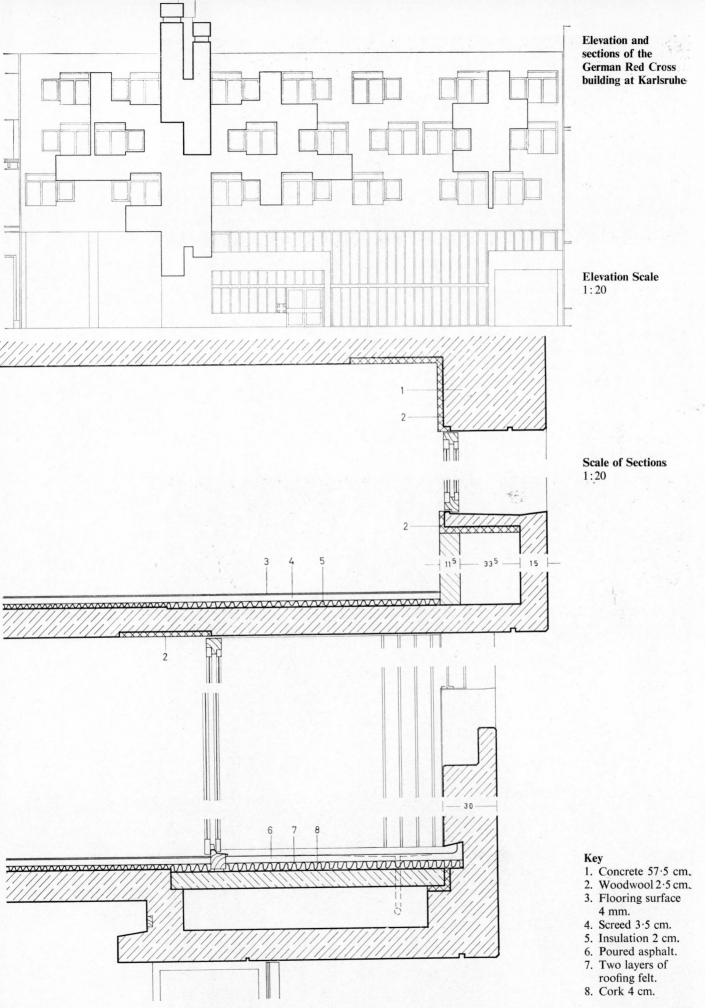

Elevation and
sections of the
German Red Cross
building at Karlsruhe

Elevation Scale
1:20

Scale of Sections
1:20

1

2

2

3 4 5

11⁵ 33⁵ 15

2

30

6 7 8

Key
1. Concrete 57·5 cm.
2. Woodwool 2·5 cm.
3. Flooring surface
   4 mm.
4. Screed 3·5 cm.
5. Insulation 2 cm.
6. Poured asphalt.
7. Two layers of
   roofing felt.
8. Cork 4 cm.

105

The Musashino University of Art, Tokyo (1964)

Architects: Yoshinobu Ashihara & Associates, Tokyo
*(Photographs on this page and on the page opposite by
Fotos Retoria/Y. Futagawa, Tokyo)*

Concrete of grey Portland cement. Mix: Cement, river sand
and river gravel in the proportions of $1 : 2 \cdot 09 : 2 \cdot 96$. The
formwork was constructed of plywood with diagonal boards
of Alaska cedar mounted on it, and was treated with mould
oil. No joints were filled, and the cast concrete received no
surface treatment.

The herring-bone pattern called for the greatest care when
the formwork was struck. To assist the process, the flanks of
the buttresses pictured in side-view below are slightly bevelled.

One of the end walls of
the Musashino Art
building

(*Below*) **Elevation and sections of balustrades on the main wall**
Scale: Elevation 1:20. Sections 1:5.
a = Tie-holes.

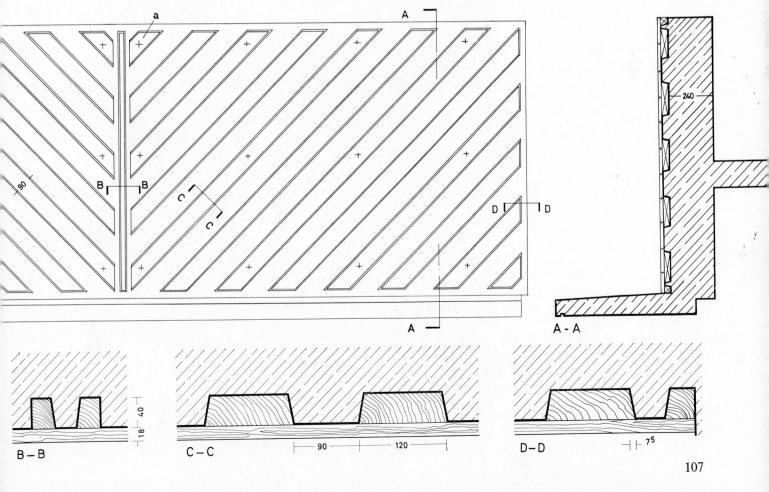

The Lichtenberg farm near Landsberg-am-Lech, Bavaria (1962)

Architect: Franz Kiessling, Munich

Associates: Hansjörg Gottlieb, Adolf Liebich and Walter Blümel

(*All photographs of the farm taken by the architect, Franz Kiessling*)

The sheds have 22 cm-thick concrete walls with a ventilated airspace at the rear, lined with 5 cm woodwool boards. The rear walls of perforated brick, one brick deep, faced with wall-tiles set in cement mortar.

The formwork was constructed of unwrought, oiled boards of varying widths, all joints being left unstopped.

The architect comments: "At a number of points the concrete leaked. This would have been prevented had we used tongued and grooved boarding. Some of the building components were shuttered with new boards. This led to variations in the colour of the concrete surfaces which I expect will tone down in the course of time."

The walls of the hay-loft (*above*) are built of round logs 9 to 11 cm in diameter

Calf-sheds with open yards running out of them. In the background, another view of the hay-loft

108

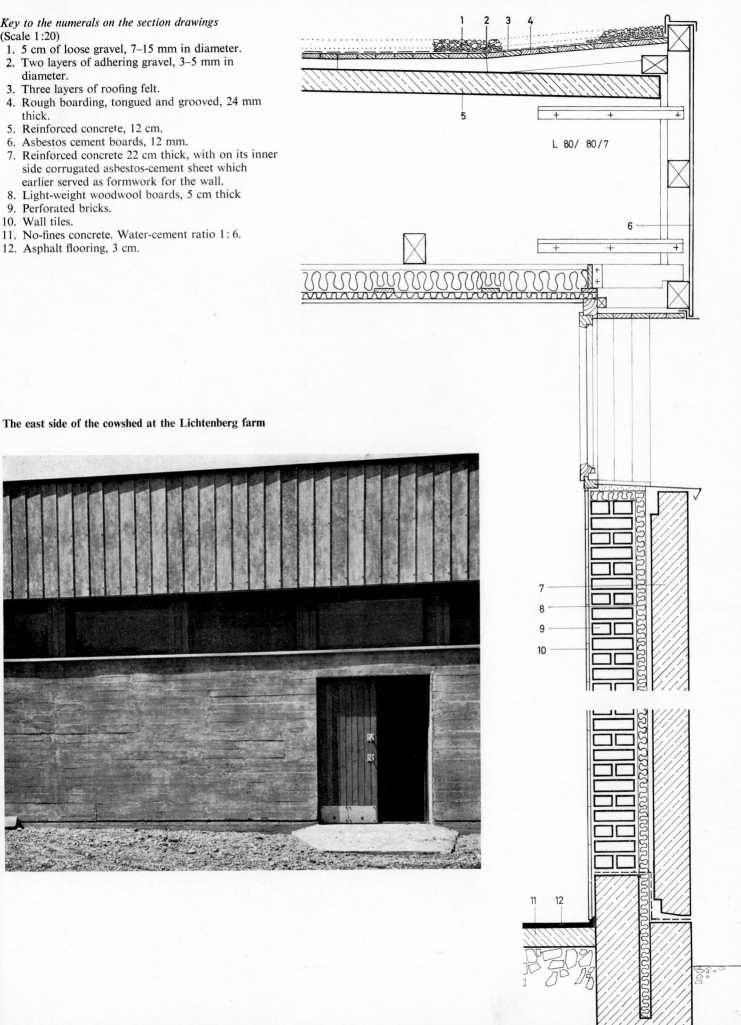

1. 5 cm of loose gravel, 7–15 mm in diameter.
2. Two layers of adhering gravel, 3–5 mm in diameter.
3. Three layers of roofing felt.
4. Rough boarding, tongued and grooved, 24 mm thick.
5. Reinforced concrete, 12 cm.
6. Asbestos cement boards, 12 mm.
7. Reinforced concrete 22 cm thick, with on its inner side corrugated asbestos-cement sheet which earlier served as formwork for the wall.
8. Light-weight woodwool boards, 5 cm thick
9. Perforated bricks.
10. Wall tiles.
11. No-fines concrete. Water-cement ratio 1:6.
12. Asphalt flooring, 3 cm.

The east side of the cowshed at the Lichtenberg farm

L 80/ 80/7

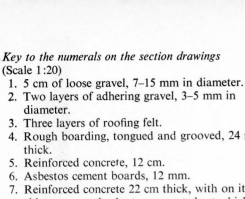

The Library of the Gakushuin University, Tokyo (1963)

Architects: Kunio Mayekawa & Associates, Tokyo

(*Photograph by Yoshio Watanabe, Tokyo*)

Noteworthy in the elevation of the Library is the patterning of the formwork in which the walls were cast so as to produce panels relatively small in size compared with the large areas of visual concrete forming the walls themselves. The effect produced is almost that of masonry.

The concrete mix per cubic metre was: 327 kg Portland cement, 692 kg sand, 1,139 kg gravel aggregate. The form-work panels were constructed of wrought boarding, 60 × 180 cm (about 2 ft × 6 ft). The walls themselves (20 cm of concrete with a 10 cm air cavity behind) have inside boarding of plywood or plasterboard. The concrete surface was coated with a transparent plastic emulsion.

Scale of Section (*below*) 1:50

Key:

1. Reinforced concrete 20 cm.
2. Air cavity 10 cm.
3. Plywood panelling.
4. Plasterboard, 7 mm thick, on batten framing covered with fabric and painted.
5. Convector-type air-heater.
6. Ribbed ceiling of reinforced concrete.

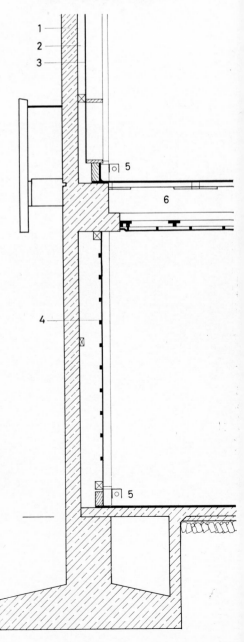

Administrative Headquarters of the Kagawa Prefecture at Takamatsu, Island of Shikoku, Japan (1958)

Architect: Kenzo Tange, Tokyo
Associates: Takashi Asada, Kofi Kamiya and Taneo Oki
Structural Engineer: Yoshikatsu Tsuboi

(*Photograph by F. Murasawa/Shokokusha, Tokyo*)

The grand formal tradition of Japanese building in wood is known the world over. The spirit of this tradition is clearly reflected in this fine building in visual concrete.

A block of office buildings—'Zur Palme'—in Zürich (1964)

Architects: Max Ernst Haefeli, Prof. Dr.h.c. Werner M.
Moser and Prof. Dr.h.c. Rudolf Steiger, Zürich
Associate: André M. Studer.

(*Photographs on this page and the page opposite by Walter Binder, Zürich*)

Sixteen floors of office space are supported on eight heavy columns rising from a two-storey podium containing shops. Circular ramps lead up to the parking areas situated on the roof of this podium.

The formwork for the visual concrete surfaces of the podium, the circular ramps and the load-bearing members on all the floors of the main building was of wrought boards 8–12 cm wide, butt-jointed, not treated with mould oil. All surfaces exposed to the weather were given a coat of silicone paint; all prefabricated elements of the façade of the main building were treated with emulsion paint.

Heat insulation was achieved by means of 3 cm of cork, applied after concreting was completed and then given a plaster finish.

A ramp leading up to one of the parking areas of the 'Palm Tree' office building in Zürich
The ramp is shown curving round one of the eight load-bearing
pillars of the main structure. Above the roof of the podium
these pillars support the reinforced-concrete platform from which
the sixteen upper storeys rise.

Chapter 5

Prefabricated building components in visual concrete

Dates

1824   Joseph Aspdin (1778–1855), of Leeds, England, applies for a patent for his method of making Portland cement.

1837   The English concrete manufacturer John Bazley White erects a country house near Swanscombe, in Kent, entirely of concrete.

1849 (approx.)   Joseph Monier (1823–1906), a gardener in Paris, hits on the idea of reinforcing concrete flower tubs with wire mesh. Takes out his first patent in 1867.

1855   François Coignet in England takes out Patent No. 2659, in which the principle of building in reinforced concrete is formulated for the first time.

Millard House, Pasadena, California (1922–1923)

Architect: Frank Lloyd Wright

(*Photograph by Ezra Stoller Associates, Rye, N.Y.*)

In his well-known *Testament*, Frank Lloyd Wright states that this was the first
house ever to be made of concrete blocks, in which the method of construction
with 'building boxes' that he had invented some years previously was applied.
It consisted of a hollow wall of three-inch-thick concrete blocks reinforced
cross-wise in the joints with thin cement mortar.

The Church of St. Matthew at Pforzheim, West Germany (1953)
Architect: Professor Dr. h. c. Egon Eiermann, Karlsruhe
(*Photograph by Eberhard Troeger, Hamburg-Grossflottbek*)

The walls between the concrete pillars are composed of pre-cast concrete
elements, square in shape but each having in its centre an octagonal opening
glazed in different colours of glass.

A block of flats in the Freimann suburb of Munich (1963)

Architect: Kurt Ackermann, Munich
Associate: Peter Jaeger

Reinforced-concrete frame construction. The mix was
300 kg of grey cement per cubic metre, with aggregate of
washed gravel grading: 0–7 mm, 60%; 7–30 mm, 40%.
Plasticizer was added, 1·2 kg to the cubic metre. No surface
treatment.

The balconies were constructed of prefabricated elements.
The formwork for them was of planed boards, tongued and
grooved, 30 mm thick and 8 cm wide, laid in staggered
lengths so that the joints of adjacent layers of board did not
coincide. New boards were aged with cement grout, then
used five times each. At all points where new formwork had
to be joined to completed lifts, foamed plastic strips were used
to plug the joints. Spacer ties were 6 mm steel rods, threaded
through plastic pipes.

After striking, the concrete was protected by plastic film
against dripping grout.

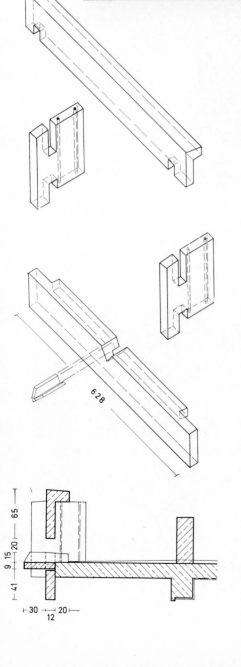

A ceramics factory at Caserta, near Naples, Italy (1960–1963)

Architects: Dr. Luigi Figini and Gino Pollini, Milan

(*Photographs by Clari, Milan*)

The buildings are of pre-fabricated concrete elements cast in metal formwork. External treatment was with transparent silicone wash; internal treatment with opaque white paint.

For the roofing of the factory bays, prefabricated concrete elements of V-shaped section were used—25 m long, prestressed and steam-cured.

The Administration Building

The roof of one of the bays of the factory

The Administration Building of the Emhart Corporation, Bloomfield, Connecticut, U.S.A. (1963)

Architects: Skidmore, Owings and Merrill, New York
(*Photograph by Ezra Stoller Associates, Rye, N.Y.*)

The entire building rests on a slab supported on 6 m columns laid out on a 12·8 m grid. Both the floor slab and the columns were cast *in situ*.

Also resting on the slab, but well forward of the window-line, are a number of square columns, 2·50 m high and 56 × 56 cm in cross-section, which carry on steel bearings the light roof construction. All first-floor edge-beams and columns are prefabricated elements of exposed concrete, each 6·5 cm thick.

The formwork was of flat, steam-formed plywood, the joints chamfered and stopped with putty. The supports of the building were cast in two lifts—first the piers, then the four diagonally cantilevered ribs with the floor slab resting on them.

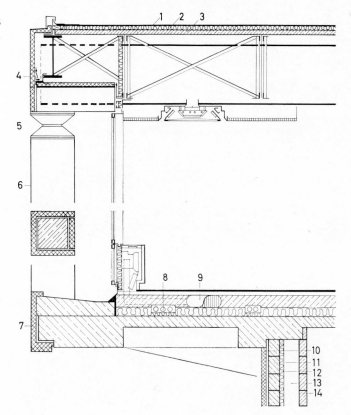

Section through the upper floor. Scale 1:50.
Key:
1. Mineral roofing felt.
2. Insulation, 5 cm.
3. Reinforced concrete slab, 5 cm.
4. Cornice facing of artificial stone, 100 × 6·5 cm.
5. Steel coupling, clad with stainless steel.
6. Column of reinforced concrete, 20 × 20 cm in section, clad with 6·5 cm-thick panels of artificial stone.
7. String course of artificial stone, 73 × 6·5 cm.
8. Duct for pipework.
9. Air-conditioning duct.
10. Artificial stone cladding, 6·5 cm thick.
11. Concrete block wall, 15 cm thick.
12. Insulating material, 5 cm.
13. Intermediate pillar of reinforced concrete.
14. Concrete block wall, 10 cm thick.

The Memorial Museum seen from the south

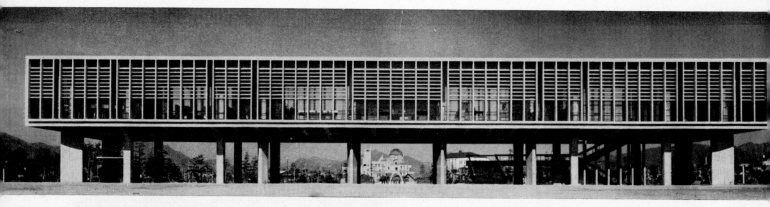

The Peace Centre at Hiroshima, Japan (1950–1955)

Architect: Kenzo Tange, Tokyo
Associates: Takashi Asada and Yukio Otani
Structural Engineer: Kiyoo Matsushita
(*Photographs by Ishimoto, Tokyo*)

The Museum building stands on a small island in a park. Through the archway can be seen buildings left unrepaired after the nuclear explosion over the city in 1945.

The Memorial Museum seen from the south

The IBM Research Laboratory at La Gaude, France (1960–1962)

Architects: Professor Marcel Breuer and Associates,
New York
Associate: Robert F. Gatje.

The building is located in the very dry region north of Nice. Its façade was assembled of prefabricated concrete elements with deep profiles to act as protection against the sun. The tapered sections of these elements greatly assisted striking of the formwork.

The concrete used was made of Portland cement and local aggregate, cast in the usual boarded shuttering and after-treated with a transparent water-repellent wash.

The grid system employed was as follows: supports, 12 × 12 m; window centres 1·80 m; internal room heights 3 m; floor-to-floor depth, 4·50 m; depth of precast elements on the elevation, 0·90 m.

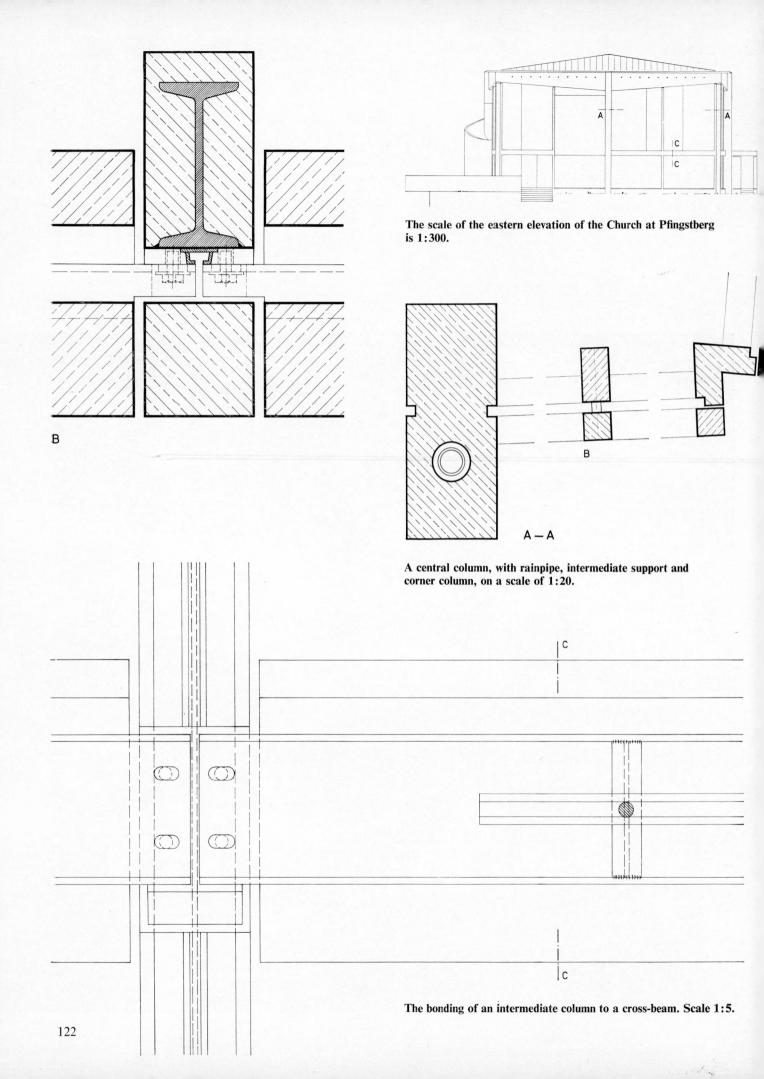

The scale of the eastern elevation of the Church at Pfingstberg is 1:300.

A — A

A central column, with rainpipe, intermediate support and corner column, on a scale of 1:20.

The bonding of an intermediate column to a cross-beam. Scale 1:5.

B

C

C

122

A south-east view of the Church, with the curving shape of the stairway to the choir loft standing out in strong relief against the huge glazed front wall of the Church

The Church at Pfingstberg, near Mannheim, West Germany (1963)

Architect: Carlfried Mutschler, Mannheim
Associate: Jürgen Bredow

(*Photograph by Robert Häusser, Mannheim-Käfertal*)

The Church itself, together with the adjoining Rectory and Community Centre, are all in visual concrete.

The mix was B300 with 270 kg grey Portland cement per cubic metre of concrete. The aggregate and sand specified was of extra-high quality, graded as follows: 0–3 mm, 32%; 3–7 mm, 18%; 7–15 mm, 25%; 15–30 mm, 25%. The surface was left untreated after striking of the formwork.

The vertical pillars of the Church and the cross-ties of the glazed walls (10 mm plate-glass) are composed of heavily-reinforced prefabricated concrete elements cast as a double-thickness component.

Formwork of 10 cm-wide boards, planed and waxed, tongued and grooved. The corner joints sealed. Since building took place in the summer, the formwork was kept well sprayed with water.

Heat-insulating boards of expanded polystyrene were fitted in the sacristy, one side of the boards being secured with spacers to the reinforcement and the other protected by expanded metal.

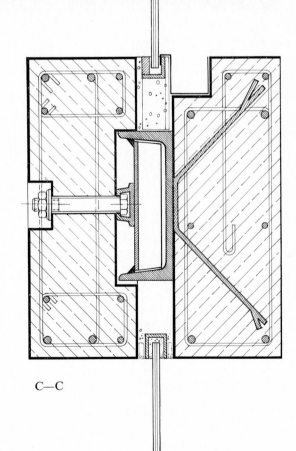

C—C

123

View from the south-east

Scale of section 1:20
Scale of detail 1:5

Key:
1. Wall of *in situ* concrete, 20 cm.
2. Mineral wool, 50 mm.
3. Air cavity, 20 mm.
4. Precast concrete slab, 8 cm.
5. Synthetic putty.

The Protestant Church at Bergshamra, Sweden (1963)

Architect: Georg Varhelyi, Stockholm

The Church building stands on a slope, and the Church itself was built over the Community Hall (*seen at ground level in the photograph above*).

The great variations in temperature to be expected in Sweden made it undesirable for the outside walls to be cast as a monolithic slab, and the following sandwich type of construction was used instead: An outer skin 8 cm thick consisting of ten-foot-high (3 m) prefabricated slabs was fixed with stainless steel cramps to an inner shell of *in situ* concrete. When the slabs had been accurately aligned, mortar was forced into their anchor cavities. Polystyrene strips were used as sealants against air penetration. The space between the two skins of the outer walls consisted of 20 mm air cavity and 50 mm mineral wool. Joints between the elements of the outer shell were sealed with Thiokol. The concrete mix for the outer skin was B300; for the inner skin B250. Grey Portland cement, with sand and gravel up to 16 mm in diameter.

The formwork used was of unwrought boards 25 mm thick. Board width was, as to nine-tenths of the boards used, 10 cm; but for the remaining tenth, only 5 cm.

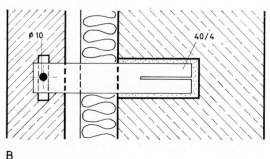

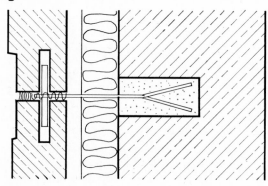

124

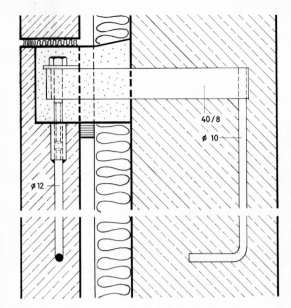

A

40 / 8

∅ 10

∅ 12

An inner wall of the Church at Bergshamra.

Detail of the east wall
of the Church building

**The main Administration Block of the Philips plant
at Düsseldorf (1960)**

Architect: Professor Paul Schneider-Esleben, Düsseldorf

(*Photograph by Inge Goertz-Bauer, Düsseldorf*)

The framework of the building was cast in *in situ* concrete
with suspended spandrels of washed concrete. Grey
blast-furnace cement and river pebbles were used for the
concrete.

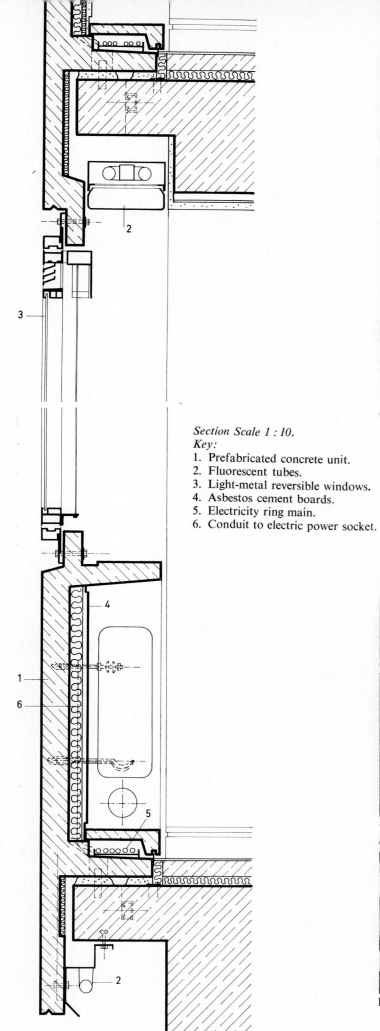

Assembly in progress at the Philips plant at Düsseldorf
(*Photograph by Helma Hermes, Düsseldorf*)

Section Scale 1 : 10.
Key:
1. Prefabricated concrete unit.
2. Fluorescent tubes.
3. Light-metal reversible windows.
4. Asbestos cement boards.
5. Electricity ring main.
6. Conduit to electric power socket.

Interior view of wall element

A primary-secondary school at Haigerloch, near Tübingen in Baden-Württemberg, West Germany (1964–1965)

Architects: Günter Behnisch, H. Bidlingmaier, M. Sabatke and H. J. Wessel, Stuttgart

The outer walls of the School are faced with reinforced-concrete spandrels and wall slabs, with cast-in heat insulation of 4 cm expanded polystyrene. No damp course. Silicone treatment inside and out.

The prefabricated wall units were made of vibrated concrete made with grey Portland cement, and with sand and aggregate in the following proportions: 0–3 mm, 28%; 3–7 mm, 25%; 7–15 mm, 27%; 15–25 mm, 20%. The water/cement ratio was 0·35. No special admixtures.

The formwork for the spandrels of the façades was of lacquered plywood treated with mould oil, the joints being stopped with a sealing gasket and a two-ingredient mastic. The columns, beams, interior walls and soffits were cast in light-grey Portland cement, with river gravel in the proportion of 1:3·5 down to 1:7·5 according to the quality and strength of the concrete required in varying locations. The water/cement ratio was 0·5, and no special admixtures were used.

The partition-walls of the classrooms were cast from vertical steel moulds which had two arms at right angles to the principal member. Horizontal steel formwork for the ceiling slabs. The individual sheets were polished and treated with mould oil. Formwork for the columns and beams was of blockboard, surface-sealed with plastic and then oiled.

It was found that components stored for a length of time under cover in the factory acquired a more uniform colour. For this reason, it is suggested that components stored in the open should always be covered with metal foil.

When assembly work had to continue in rain or snow, mortar newly placed in the joints of the roof-slabs repeatedly caused white streaks to appear on components lying beneath, which were very difficult to eradicate. Some of the components cast in vertical formwork occasionally showed dark stains for which no explanation could be found.

Interior view of a classroom in the School at Haigerloch
(*Photographs on this and the facing page by Gottfried Planck, Stuttgart*)

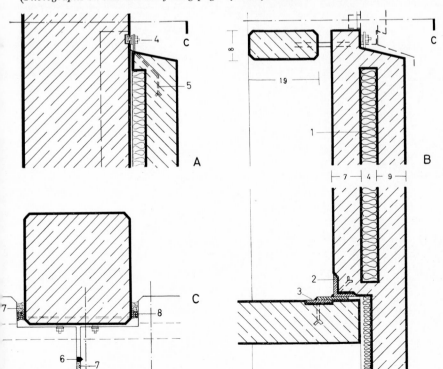

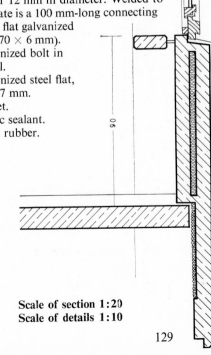

Scale of section 1:20
Scale of details 1:10

129

A secondary school at Furtwangen in the Black Forest, West Germany (1963–1964)

Architects: Günter Behnisch, L. Seidel, K.-H. Weber, Stuttgart

Associate: P. Schirm

(*Photographs by Gottfried Planck, Stuttgart*)

All windows facing south are protected by aluminium baffles against excessive sunlight, which can cause trouble even in mid-winter

Each wall unit contains, side by side, two double-glazed casements pivoting round a vertical axis, with two ventilators above. The long horizontal slots in the spandrels are for ventilation

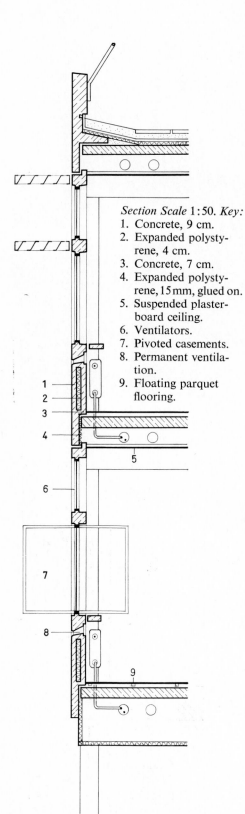

Section Scale 1:50. Key:
1. Concrete, 9 cm.
2. Expanded polystyrene, 4 cm.
3. Concrete, 7 cm.
4. Expanded polystyrene, 15 mm, glued on.
5. Suspended plasterboard ceiling.
6. Ventilators.
7. Pivoted casements.
8. Permanent ventilation.
9. Floating parquet flooring.

130

The Chemical Institute Buildings, Technical University of Karlsruhe, West Germany (1964–1966)

Design, Planning and Supervision:
The University Building Department, Karlsruhe

Collaborating Architects: Schmitt and Kasimir

(*Photographs by Thilo Mechav, Karlsruhe*)

The complete building was assembled from prefabricated elements. Concrete mix: Portland cement with carefully selected river gravel aggregate, grading 0–30 mm. Water-cement ratio: 0·45 to 0·5.

All components were cast in steel formwork at the factory. Later, in an attempt to minimize unsightly streaks and variations in colour, all visible surfaces were treated with a transparent varnish.

Fire-escape balconies run round the façades of all the Institute buildings

This photograph was taken before the final coat of varnish had been applied to the visual concrete of the balcony panels.

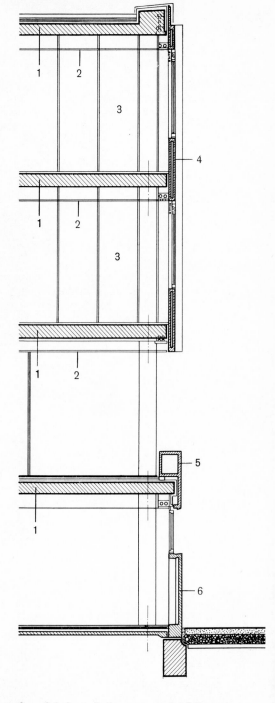

The Teachers' Training College and State College of Sport, Ludwigsburg, near Stuttgart (1963–1966)

Architect: Professor Erwin Heinle, Stuttgart
Associate: Helmut Wiedmann

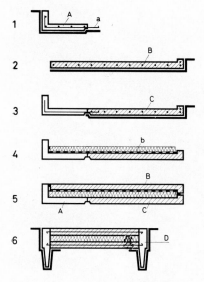

Detail of constructional elements used on the façades:

1. Casting of Outer Skin A (in white concrete).
2. Casting of Inner Skin B (in grey concrete).
3. Casting of junction between Outer Skin C and Element A (grey concrete).
4. Heat insulation of foam glass glued to Outer Skin.
5. Inner Skin assembled.
6. Uprights D (white concrete) cast on to the assembled panel.
   a Joint-sheathing.
   b Glued joint.

A combination of prefabricated elements and of floors cast on the site, using the lift-slab system. Columns of visual concrete 17·5 metres (about 58 ft) high and 50 × 50 cm in section rise through four floors of the building, smooth-faced on all sides but with sharp arrises. They were cast vertically in storey-high sections, then laid horizontal, jointed together with synthetic resin and "post-tensioned" (*i.e.*, locked by a steel member running the full height of the column). The completed column was then picked up by crane and placed upright into prepared holes in the foundations (grid: 9 × 9 m).

After the erection and careful adjustment of the columns, the four floor slabs were cast one on top of the other on the ground, being kept separated by a special spray. Each slab has an area of some 850 square metres and weighs nearly 650 tons. Hydraulic lifting gear was mounted on the top of each column, and a lifting speed of about 2 metres per hour was attained. This method of ground-casting cut construction time, and reduced both the number of skilled workmen and the quantity of formwork required.

A corner of the Teaching Wing of the Ludwigsburg Hochschule

A close-up view of the concrete components assembled to form the façade shows an "elevation unit" rising two storeys high, and comprising two slender uprights with 2·25 metre-wide spandrel slabs between them. Every unit is bolted to and suspended from the roof slab (having been reinforced by means of a special "steel corset" during assembly).

The 15 mm-wide vertical joints between the units were made rainproof with neoprene tubing and a Thiokol sealant applied with the gun.

Waterproofing of all outer surfaces was achieved by spraying with a silicone-based solution. Interior surfaces in visual concrete were made dirt-resistant by impregnation with a transparent wash.

Internal partitions are of prefabricated elements of reinforced concrete, smooth-finished and exposed on both sides, about six inches thick, four feet wide and 12 feet high (15 cm × 1·20 m × 3·60 m). All vertical joints were filled with hollow gaskets of PVC.

Clear ceiling height on each floor is 3·60 metres.

Scale of section on preceding page, 1:100
Key:
1. Floor-slabs lifted into position by crane.
2. Suspended ceiling, serving as a heat radiator.
3. Prefabricated partition-walls in concrete, movable.
4. Prefabricated element on façade.
5. Prefabricated spandrel.
6. Prefabricated wall element.

The College Gymnasium
(*Photograph by Gottfried Planck, Stuttgart*)

Six supporting frames at 18 m centres. Columns of prefabricated components. The horizontal members of prestressed hollow beams cast *in situ*. Roof and all elements of the elevations assembled from prefabricated components.

A group of buildings at the University of Marburg, Province of Hesse, West Germany (1964–1966)

Design: State Office for New University Buildings, Marburg
Leader: Kurt Schneider (State Director of Construction)
Associates: Helmut Spieker, Dieter Biel, Gerhard Haberle, Hilger Schallehn

The Faculties of Science and Medicine of the University are in process of moving from the old University area in the City to a new 250-hectare (about 620 acres) campus lying to the east of the town. There it is proposed to build to the extent of some 2·5 million cubic metres—an operation for which a fully-prefabricated assembly system of great flexibility has been developed.

The principle on which this system works is as follows: A reinforced concrete frame is built up bit by bit from a number of interchangeable "building units", each consisting of four prefabricated columns tied together by a prefabricated floor slab. Successive building units of this type are placed side by side over the full area of the ground floor of the building; and the second and subsequent storeys, similarly consisting of individual prefabricated building units, are laid on top, up to the full designed height of the building.

It will be seen that there are always a number of points in the interior of the grid at which four building units adjoin, and that at these points four columns will be closely grouped. On the exterior of the grid, where at most two building units can lie side by side, only two columns will be closely adjacent to one another (*two examples of this latter arrangement can be seen in the photographs to the left and on the facing page*). At all points where columns adjoin, whether in groups of two or four, space is left for the vertical channeling of services or for the horizontal insertion of concrete rainwater spouts.

External and internal walls have frames into which fit interchangeable panel infills. All steel sections are galvanized, with stressed plastic skins of neoprene. All elements jointed by means of steel cramps. The balustrade units consist of twin skins of 6 mm glazed asbestos-cement panels, with plastic-reinforced mineral-wool infilling.

The balconies serve also as escape routes in case of fire
(*Photograph by H. Bickon*)

Twin columns, balustrading and a rainwater spout in the
prefabricated Faculty Buildings at Marburg

Chapter 6

Arches, vaults and shells in visual concrete

Extract from the French master patent issued to Joseph Monier

Inventor's Patent: Period of Validity, 15 years

Date: July 16th, 1867   No. 77,165

Nature of Product: A movable tub-container of iron and cement, for use in making large flower-pots.

Applicant: Joseph Monier

"The container can be of any size or type—square, round or oval, etc., and with or without openings. The method of manufacture is always the same. With round or square iron bars I make a trellis-work (*grillage*) to the desired shape (as shown in Diagrams 1–3) and cover it with any desired cement such as Portland, Vassy, etc. to a thickness of 1 to 4 cm, depending on the size of the object in question."

(Quoted in *Vom Caementum zum Spannbeton*, Bauverlag Wiesbaden-Berlin, 1964).

The bridge at Langwies on the railway line from Chur to Arosa, Switzerland (1912–1914)

Structural Engineer: H. Schürch
Construction: Ed. Züblin AG

(*Photograph by C. Brandt, Arosa-Archiv Züblin*)

This 50-year-old bridge could hardly be bettered for elegance today. In its own time it ranked among the biggest and boldest reinforced-concrete bridges in the world.

The span of the arch is 100 m, its camber 42 m, and it rises to a height of about 70 m (nearly 230 ft) above the valley floor.

The building of the great arch called for the construction of an ingenious arrangement of fan-shaped wooden scaffolding supported on a tower-like reinforced-concrete base which was demolished after completion.

**The bridge across the Schwandbach, near
Schwarzenburg, Canton Berne (1933)**

Structural Engineer: Robert Maillart

*Photograph from "Raum, Zeit, Architektur" by S. Giedion
(Otto Maier Verlag, Ravensburg, 1965)*

Span Widths of some Reinforced-Concrete Bridges

1893   The bridge over the Danube near Munderkingen
       —50 m (162 ft).

1904   The bridge across the Isar at München-
       Grünwald—70 m (227 ft).

1914   The Langwies bridge pictured on the preceding
       page—100 m (325 ft).

**The bridge follows the curve of the road running across it.
The span of its arch is 37·4 m (about 122 feet)**

A road bridge over the Glems at Schwieberdingen, near Stuttgart (1960–1961)

Design: Wayss & Freytag KG (Stuttgart branch) and Professor Wilhelm Tiedje, Stuttgart

Construction: Wayss & Freytag KG and Karl Kübler AG (in collaboration)

(*Photographs by Archiv Wayss & Freytag, Frankfurt-am-Main*)

Span 114 m; camber 28·75 m; maximum height above ground 38 m. To preserve the clean lines of the pre-stressed concrete, the formwork used was of three-ply boards faced on all sides with plastic and with their corners left unshielded. Most of the boards were used repeatedly, some of them up to seven times. The panels were 50 cm wide, preferred lengths being 250 cm and 500 cm with intermediate sizes at 50 cm intervals.

Special attention was paid to the maintenance of operations at a continuous tempo throughout.

After careful experiment, the most successful mixes were established as follows: *For the pillar walls* (B300): 300 kg cement; 60 kg trass (pozzalana); 0·6 kg Tricosal admixture. Aggregate was graded as follows: 0–3 mm, 40%; 7–15 mm, 15%; 15–30 mm, 15%; 30–70 mm, 30%. *For the boxed arch and superstructure* (B450): 320 kg cement; 40 kg trass; 0·48 Tricosal. Aggregate was graded: 0–3 mm, 38%; 3–7 mm, 12%; 7–15 mm, 20%; 15–30 mm, 30%.

The concreting of the lifts followed one another at intervals of 1½–3 hours so that the top layers of the preceding lift could be revibrated with the new one without the use of retarders being necessary.

A recent view of the interior
(*Photograph by Tomasz Olszewski/Zaiks, Wroclaw*)

The Centenary Hall, Breslau (Wroclaw) (1911–1912)

Architect: Max Berg
Construction: Dyckerhoff and Widmann KG

Dome construction of reinforced concrete, covering an area of 5300 square metres (over 57,000 sq ft). Dome diameter 65 m, and height 42 m.

The Centenary Hall has served as a model for all later large buildings in visual concrete. Its span of over 213 ft clearly demonstrated at that early date the structural potential of the new material, at the same time as the clean sweep of its vaults and arches suggested its aesthetic possibilities.

The Great Dome of the Centenary Hall in Breslau
(*Photograph by Dyckerhoff & Widmann KG, Archiv*)

Two Sports Halls built for the Tokyo Olympics
(1962–1964)

Architects: Kenzo Tange, Urbanist and Architect Team,
        Tokyo
Constructional Engineers: Yoshikatsu Tsuboi, Mamoru
                       Kawaguchi and Shigeru Kawamata

(*Photographs on this page and the page opposite by Osamu Murai,
  Tokyo*)

Opposite: **A view from the Garden Court of the Swimming Baths
towards the smaller Hall**

The larger of the two Halls in Kenzo Tange's Olympic
complex seats 15,000 people, and contains swimming pools
which can also be converted into ice rinks. The smaller Hall
seats 4,000 spectators.

    The building linking the two Halls contain the Administra-
tion block and a restaurant.

The smaller Sports Hall (Palazzetto dello Sport) in Rome (1957)

Architect: Annibale Vitellozzi, Rome
Construction Engineer: Dr. Pier Luigi Nervi, Rome
(*Photographs on this page and the page opposite by G. Gherardi & A. Fiorelli, Rome*)

The stresses of the shallow dome are transferred to 36 supports. The covered area extends to 4,766 square metres (nearly 1¼ acres) and seats 5000 spectators.

The dome itself is supported by a network of inclined ribs in heavily reinforced concrete. The triangular and rhomboid prefabricated components in reinforced concrete which served as 'lost' (*i.e.*, unrecovered) formwork were left to form an integral part of the building.

A closer view of the inclined concrete ribs which support the dome

An evening view under the dome of the Palazzetto dello Sport

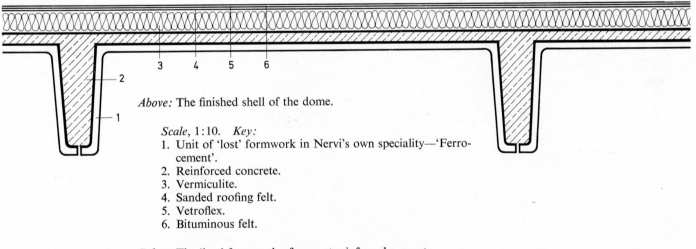

Above: The finished shell of the dome.

Scale, 1:10.   *Key:*
1. Unit of 'lost' formwork in Nervi's own speciality—'Ferro-cement'.
2. Reinforced concrete.
3. Vermiculite.
4. Sanded roofing felt.
5. Vetroflex.
6. Bituminous felt.

Below: The 'lost' formwork of precast reinforced concrete.

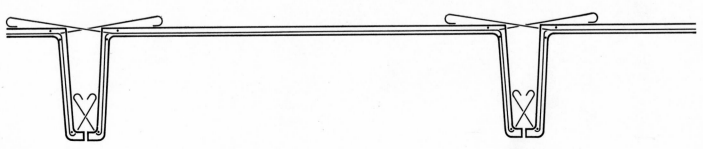

The stands at the racecourse of La Zarzuela, near Madrid (1935)

Architects: C. Arniches and L. Domínguez, Madrid
Constuction Engineer: Eduardo Torroja

The roof of every stand consists of a series of 'single-shell hyperboloids', each of them cantilevering 12·57 m (41 feet) and tied at the back by tension members.

A closer view of a section of the roof on one of the stands

"Los Manantiales" Restaurant at Xochimilco, near Mexico City (1957–1958)

Architects: Joaquín and Fernando Alvárez Ordóñez, Mexico City
Construction Engineer: Félix Candela, Mexico City

Eight groined vaults, made up of four pairs of interpenetrating hyperbolic paraboloids. Maximum diameter 42·42 metres (139 ft). Camber of the arches 9·93 metres. Height at centre of vaults 5·84 m (19 ft).

"Félix Candela, an architect and engineer of Spanish origin, came to Mexico in 1939 and has practised there ever since. He has become internationally known for his shell structures, especially those based on the hyperbolic paraboloid, with which he has covered in spaces of all sizes, creating buildings that are economical to erect, flexible in use and sculpturally exciting"—J. M. Richards, Architectural Correspondent of *The Times*.

One of the eight supports of the restaurant building

The Hall of Cement at the National Exhibition, Zürich (1939)

Architect: Hans Leuzinger, Zollikon
Construction Engineer: Robert Maillart

Photograph taken from "Raum, Zeit, Architektur" by S. Giedion (Otto Maier Verlag, Ravensburg, 1965)

The building is composed of two parabolic shells, one of them having a conical taper along its length. The bridge acts as a tension member.

The concrete was applied with a gun. The thickness of the shell is 6 cm.

Pavilion for Research into Cosmic Radiation at the University of Mexico City (1951)

Architect: Jorge Gonzáles Reyna, Mexico City
Construction Engineer: Félix Candela, Mexico City

The shell of the building had to be of minimum thickness so that cosmic rays could be measured in the interior of the building.

Two hyperbolic paraboloids are linked together along a main parabola. Diameter of arch 12 m; its length 12·75 m. The shell itself could only be 15 mm thick at the apex, but increases in thickness towards the abutments to 5 cm. Reinforcement of 3 mm steel webbing with 10 cm × 10 cm mesh.

Despite the thinness and steep pitch of the formwork, concreting was completed in about eight hours. The proportions of the mix were 1:2:3. Size of aggregate went up to 6 mm.

Formwork of 12 mm tongued and grooved boards nailed to wooden battens 5 cm by 10 cm. A number of straight lines were drawn along the surface of the paraboloid, and the formwork was constructed to follow these lines.

Chapter 7

Architectural sculpture in visual concrete

All great epochs of building have been characterized by a large measure of integration between the visual arts. In these epochs, painting, sculpture and architecture have often come close to being fused into an inseparable unity. Although today these arts have tended to grow apart and to go their own ways, the urge towards reintegration is still powerful.

For visual concrete it may be claimed that in it (and perhaps in it alone) lie the possibilities of a final synthesis of two of the visual arts—architecture and sculpture. But the perception of these possibilities will only be achieved if architect and sculptor embark together on a common intellectual adventure. For both, the avenues opened up by this new material are unexplored, and every experiment still represents a step into the more or less unknown.

But for both also there beckons an exciting common goal. It is nothing less than achievement at last of a perfect and indissoluble synthesis of sculpture and architecture into a single art form.

Wall sculpture at the entrance to a dwelling unit, Marseilles (1947–1952)

Architect: Le Corbusier

(*Photograph by Lucien Hervé, Paris*)

The "Modulor" which Le Corbusier cast in the reinforced concrete of the wall at the entrance to this block of flats is a symbol linking the building to the twin elements of the "modulor" system— Man and Measure.

Youth Centre at Ludwigshafen-am-Rhein, in the Palatinate (1963–1966)

Architect: Karl-Maria Sommer, of the State Building Office, Ludwigshafen

Sculptor: Professor Klaus Arnold, of Grein, in the Odenwald north of Heidelberg

(*Photographs by Pe Wolf, Karlsruhe*)

The sculptural elements were incorporated into the formwork, and were cast as a unit with the wall itself. The object was to identify the Centre externally, by the simplest possible means, as a gathering-place for young people.

The sculptor caused the formwork to be sawn horizontally into semi-circular disks of varying radius, which were then laid one on top of another in layers. When the formwork was struck, the concrete formed by each of the disks had assumed the shape of a large, flat, semi-circular cheese. Collectively, these "cheeses" made up the positive and negative balustrade patterns shown in the photographs on this page.

The Church of St. John at Korb, near Stuttgart (1965–1966)

Architects: Professor Hans Kemmerer and Walter Belz, Stuttgart
Sculptor: Hans Dieter Bohnet, Stuttgart

The sculpture shown is some 8 metres high (over 26 ft) and serves to direct attention to the entrance porch lying beneath. The same wall also forms the rear wall of the choir loft.

The sculptor carved the negative of his design into expanded polystyrene blocks, which were then incorporated into the formwork and concreted at the same time as the building itself.

The photograph was taken before the Church was completed, so that only the beams of the roof are visible in the background.

A close-up view of the Long Wall leading into the Memorial Hall in the churchyard of Frankfurt-am-Main/Westhausen (1962)

Architect: Günter Bock, Frankfurt-am-Main
Sculptor: Herbert Hajek, Stuttgart

(*Photograph by Abisag Tüllmann, Frankfurt-am-Main*)

A 72·5-metre long wall in visual concrete, 4·14 metres high and 37 cm thick, leads into the Memorial Hall—its dimensions in Imperial measures being about 238 ft long, 13 ft 6 in high and 14½ in thick.

On both sides of this wall there is impressed an abstract design in negative relief. Sculptor Hajek cut the positive moulds for this design with a heated wire loop from blocks of expanded polystyrene, and the moulds were then incorporated into the formwork.

154

Picasso's decorations to the Chamber of Architects, Barcelona (1961)

Architect: Xavier Busquets, Barcelona
Decoration: Pablo Picasso (Vallauris) and Carl Nesjar (Oslo).

The slabs were made of white concrete, with black pebbles as aggregate. By means of a special sand-blasting technique developed by Carl Nesjar, the outer skin of concrete was cut away so that Picasso's designs stood out in black against a white background.

Above: **Detail of the frieze running round the first floor**
(*Photograph by X. Busquets, Barcelona*)

Left: **The Foyer on the first floor**
(*Photograph by F. Catalá Roca, Barcelona*)

155

**The Theatre and Festival Hall at Ingolstadt,
Bavaria (1963–1965)**

Architects: Hardt-Waltherr Hämer and Marie-Brigitte
Hämer-Buro, Ingolstadt
Murals: Heinrich Eichmann, Zürich
(*Photograph by Helmut Bauer, Ingolstadt*)

Murals of gold-leaf were applied directly to the exposed
concrete walls of the auditorium. As varying lighting in the
hall makes the pictures lighter or darker, so they appear to
stand out from or recede into the solid wall itself.
(*For further views and details of the Theatre at Ingolstadt,
see pages 94–97*).

Chapter 8

Visual concrete in the British Isles

This Chapter presents a selection of some outstanding achievements by architects working with visual concrete in the British Isles, beginning with a technical accomplishment of astonishing virtuosity dating from just before the First World War. The Tempest Anderson Hall extension to York Museum may indeed be (as Mr. George Perkin, ARIBA, has said) 'an example of the bogus use of concrete, insofar as it is here masquerading as stone'. But the building is nevertheless of interest, apart from the high quality of the workmanship displayed, by reason of the way in which its concrete surface has weathered after nearly 60 years' exposure to the weather of the Vale of York.

Since that date, and especially since the end of the Second World War, the use of concrete as a constructional material whose surface is deliberately planned to remain visible after the building is completed has taken enormous strides; and some words which appeared in the leading American journal *ARCHITECTURAL FORUM* are as true of the British Isles today as they were of the sub-continent of which they were written some eight years ago. 'Suddenly', wrote *ARCHITECTURAL FORUM*, 'it seems as if the most striking new buildings the world over are being built of concrete, boldly and expressively exposed. Perhaps the most significant change in the use of concrete is this increasing exposure. Even the most important new monumental structures are built of raw, unfinished concrete—and with stunning effect . . . .

'Behind the growing interest in the aesthetic potentialities of concrete, there is now a reassuring body of technical knowledge built up over the past sixty years. We now know enough about the strength, the durability, and the appearance of concrete to use it with a high degree of confidence.'

The Tempest Anderson Hall Extension to York Museum (1912)

Designer: E. Ridsdale Tate
Contractor: Trussed Concrete Steel Company
(now Truscon Ltd.)

The Hall is a structural concrete extension some 79 ft long by 47 ft wide added to the Museum—a building designed in the full classical style and constructed of the local buff-coloured stone. It stands in open parkland close to the ruins of St. Mary's Abbey, in which the York Mystery Plays are performed every three years.

"It has weathered", says Mr. George Perkin, ARIBA, Editor of *Concrete Quarterly*, "with lively variations and some subtleties of colour. The pilasters, architraves and projecting parts which receive most washing down are a warm yellowish-grey. The recessed and more protected panels, by contrast, are much darker and of a brownish colour. At the base and on the plinths there are drifts of silver, merging to black, with patches of algae—this part not at all unlike the behaviour of stone . . . .

"The success of the weathering seems largely due to the mouldings, drips, projections, cornices, architraves, pilasters and (other) classical impedimenta which comprise the front. It is these which break up the surfaces and control the stains, streaks and other ravages of time to which all facing materials are heir. (No) blemish can reach any size and get out of hand; it must inevitably hit a ledge or a projection before it becomes an eyesore."

The principal "faults" to have appeared in the concrete can be seen under the three central windows, where some old ventilation grilles had to be renewed and the surrounding concrete hacked out and made good.

The point at which the new extension and the original Museum of 1829 meet. Stone to the right: concrete to the left!

The Doric Entrance Portico to the Tempest Anderson Hall

Note the astonishing precision of the horizontal and vertical formwork. The vertical boards used to form the rounded corner pilasters were no more than $2\frac{1}{4}$ inches (55 mm) wide.

(*Photographs by Kershaw Studios, Davygate, York*)

A Corner detail of the Portico

Most of the board-markings on the building as a whole are horizontal (the principal exceptions being the corner pilasters of the portico seen above). Modern practice might have preferred some attempt at articulation, perhaps with vertical formwork for the pilasters of the main façade and horizontal shuttering for the panels.

The Water-tower and Sun-room at Wappingthorn Farm, Steyning, Sussex (1929-1930)

Architect: Maxwell Ayrton, FRIBA

(*Photograph by Geoffrey Harper*)

This fine octagonal building, commanding a magnificent view over the Sussex Downs, is reached by a 5 ft wide outside stairway—the whole being cast in visual concrete, lightly and evenly bush-hammered overall. The colour of the concrete is now that of the aggregate, a warm buff heightened by spots of bright yellow lichen. The only blemishes are one or two spots of iron pyrites which have caused rust stains. Otherwise, the weathering has been (to quote Mr. J. Gilchrist Wilson, FRIBA) "most attractive and gentlemanly".

Besides the Water-Tower, other concrete buildings erected at the time include a circular dairy building and two linked silo towers. The concrete of which all the buildings were cast was of very high quality, a well-compacted mix 1:2:2 by volume, with the maximum size of aggregate $\frac{3}{8}$ in. It was cast in 2 ft 6 in lifts, the line of each lift being clearly visible on the photograph.

The late Mr. Maxwell Ayrton was, says Mr. Gilchrist Wilson, a pioneer in the use of concrete in Britain, with "a feeling and understanding for concrete that can be likened to that of a potter for his clay .... Few English architects since have used this material with greater understanding, both of its possibilities and of its limitations."

The concrete buildings at Wappingthorn were built as a model dairy farm for Sir Arthur Howard, who still lives there. "In all the forty years I have lived at Wappingthorn," Sir Arthur has written, "I have hardly spent a penny on structural repairs. I have from time to time made some alterations, and the builders have always had to spend more time and labour in the cutting-away or removal of the structure than they have in installing what was to replace it".

When a series of holes had to be drilled to provide ventilation and drainage channels at the base of the two silo towers, it set the local builders a major problem, the quality of the concrete being so much superior to anything they had encountered elsewhere.

160

Dunelm House, University of Durham, and the Kingsgate Footbridge linking the older part of the University on the Cathedral Peninsula to the newer development south of the city

Architects for the Building:
 Architects' Co-Partnership
Engineers: Ove Arup &
 Partners

(*Photographs by John Donat*)

A point of interest in the building is the extensive use made of light-weight foamed slag concrete for walls. Load-bearing external walls with a board-marked finish on both sides, and with good insulation, were required; and these had to match up with ordinary dense board-marked concrete sometimes forming part of the same ten-inch thick wall.

A view of the River Wear from Dunelm House, looking through one arm of the footbridge

The Elephant and Rhinoceros House at the London Zoo (1965)

Architects: Casson, Conder and Partners (*Partners in Charge:* Sir Hugh Casson, Neville Conder. *Associate:* Montague Turland)

Engineers: Jenkins and Potter

Contractors: John Mowlem & Co. Ltd.

Exterior view of the rhino side of the building, seen from the rhino paddock. The screens in front of the hollows between each pair of pens allow keepers an escape if animals get irritated.

The walls were cast in three lifts, with the top of the first lift coinciding with the tops of all wall openings. The timber formwork with its wedge-shaped fillets was repeatedly re-used. In the architect's drawings the precise number of fillets was specified in order to ensure evenly spaced ribbing round the curves. The concrete ribs were later roughly hacked by hand, by a single man using hammer and bolster.

Insulation is by means of one-inch polystyrene protected by an inner skin of brick, to which are fixed pale grey-blue ceramic mosaics forming cyclorama walls for the animals' dens seen from inside the building.

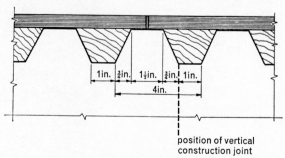

Above: **Detail of the formwork used to produce the ribbed exterior surfaces**

(*Photograph by Henk Snoek*)

162

The pavilion houses and displays four elephants and four rhinos in four paired pens, each having access to a sick-bay pen and to moated paddocks outside. The pen floors are set slightly higher than the public floor, to heighten the dramatic presentation of the huge animals in their top-lit pens.

These have no bars. Only a concrete ditch separates the animals from the public moving around in the low, dark interior with its labyrinth of concrete pillars interlacing overhead like the branches of a tropical forest.

The public viewing area (*above*) and ground plan of the Elephant House

(*Photograph by Henk Snoek*)

Apparently all elephantine curves, the pavilion is designed to within an inch of its rough-ribbed concrete walls so that all damageable objects are just out of trunk range of the largest elephant (elephants' trunks have a muscle inside which can undo almost anything).

Elephants need enormous quantities of food—about 2 cwt (over 10 kg) each a day. The ground plan of the Elephant House (*below*) shows the care with which storage facilities were planned to match these formidable demands.

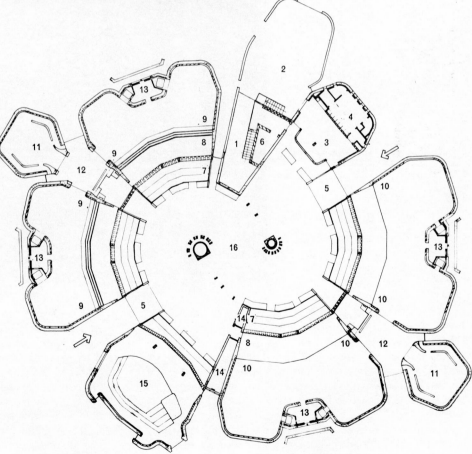

Ground Floor Plan

Key:

1-2. *Service Yard and Ramp*
3. *Staff Mess Room*
4. *Staff Lavatories*
5. *Public Entrances*
6. *Store*
7. *Public Benches*
8. *Ditches*
9. *Rhinoceros Dens*
10. *Elephant Dens*
11. *Animal Sick Bays*
12. *Den Lobbies*
13. *Drinking Troughs*
14. *Main Rising Ducts*
15. *Elephant Pool (heated)*
16. *Public Viewing Space*

New Theatre at Eton College, Windsor (1968)

Architects and Structural Engineers: A. M. Gear & Associates
Main Contractor: Taylor Woodrow Construction Ltd.

(*Photograph by S. W. Newbery*)

Eton's new theatre, a gift to the College from the Farrer Trust, overlooks the Parade Ground, standing between the School of Mechanics and the Drawing Schools (*from the portico of which it is seen above*). It seats 401 persons, and is used for lectures, films and a number of teaching purposes, as well as for drama. An older cottage adjoining the site has been converted into dressing-rooms and linked to the new Theatre.

Special problems were that the site on which the theatre stands is liable to flooding from the adjoining Thames, and that the noise of low-flying aircraft taking off from Heathrow Airport a few miles to the east is frequently appalling. The second of these problems suggested heavy construction to give a large sound-transmission loss; yet the need for bored piles to carry the loads down to firm ground made it economically desirable to keep the weight down.

The compromise adopted was to build an outer envelope of reinforced concrete, cast *in situ* 6 in. thick, with an inner leaf of brick. The roof envelope is of six-inch thick reinforced concrete covered with woodwork slabs.

The treatment of the visible surface of the outer envelope is of special interest.

Hammering the concrete ribs of Eton's new theatre

(*Photographs by David Sawtell*)

The external ribbing of the concrete was achieved by casting against timber formwork fitted with fillets 1½ inches deep, placed at four-inch centres. The projections were then Kango-hammered, not continuously but at spaced intervals on alternate sides of each rib. The wavy guide lines drawn with a template on the unhammered ribs to assist the operator can just be seen in the photograph above.

The effect was to give a subtle twist to the appearance of the projecting ribs, and an interesting character to the surface texture as a whole. The very even patterning achieved is seen in the photo-graph to the right, which shows parts of the Theatre before and after hammering.

The South Bank Art Centre, London (1952-68):
A View from the North Bank of the River Thames

Architect to the Greater London Council: Sir Hubert
   Bennett, FRIBA, FSIA
Deputy Architect: Jack Whittle
Group Leaders for Design: E. J. Blyth and Norman Engleback
Job Architect: J. W. Szymaniak
Structural Engineers: Ove Arup & Partners
   (*Partner-in-charge:* Peter Dunican)

The photograph (*by Colin Westwood*) shows Waterloo
Bridge on the left, the Royal Festival Hall on the right and
part of Shell Centre (headquarters of Shell Transport and
Trading Co.) in the background. The bold concrete outline
of the Queen Elizabeth Hall, with the Purcell Room and
Hayward Art Gallery behind it, is seen in the centre of the
photograph. The National Film Theatre is under Waterloo
Bridge: the site of the future National Theatre just beyond it.

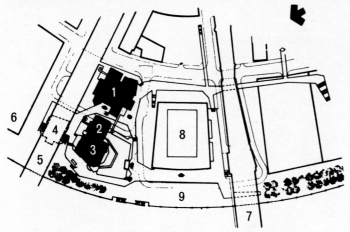

Site plan of the South Bank Arts Centre

1. *Hayward Art Gallery*
2. *Purcell Room*
3. *Queen Elizabeth Hall*
4. *National Film Theatre*
5. *Waterloo Bridge*
6. *Site of National Theatre*
7. *Hungerford Bridge*
8. *Royal Festival Hall*
9. *Riverside Walk*

The Hayward Art Gallery and Pedestrian Terrace, South Bank Arts Centre, opened by H.M. The Queen on July 1st, 1968. A view towards Big Ben and the Houses of Parliament over on the North side of the River

(*Photograph by Richard Einzig*)

The fine detail of the board-marked concrete is clearly seen. Also to be noted is the high quality of the precast cladding panels, made of crushed Cornish granite with the joints filled with polysulphide filler.

The entire area of the Centre is linked by broad pedestrian terraces running at different levels round the several buildings forming the complex.

The foyer of the Concert Halls in the South Bank Arts Centre

(*Photographs by Colin Westwood*)

A combination of pale-grey board-marked visual concrete for the columns, walls and beams, with slatted aluminium sound absorbers in the ceiling and white marble on the floor.

The Queen Elizabeth Hall is designed as a concert hall, seating 1,106 persons, while the Purcell Room (seating 372) is used for musical recitals. The two share a common foyer, seen above. They were opened by The Queen in March 1967.

The Hall was designed as a reinforced concrete box, 154 ft long by 86 ft wide by 70 ft high, with the rear of the auditorium as a propped cantilever and the walls as structural members. The roof is a 15 in. thick concrete slab on concrete box beams. The Purcell Room is a similar concrete box, but its fourth side projects as a cantilever.

Special care was taken to achieve high standards in the board-marked concrete, a member of the GLC Architects' Department undertaking an extensive European tour before work began to examine some of the outstanding results already achieved in Germany, Switzerland and Italy.

Right: **Detail of the mushroom-headed concrete column used in the buildings and external walkways of the Arts Centre**
(*Photograph by Colin Westwood*)

When the Arts Centre went out to tender, contractors were shown a sample panel of exposed concrete derived from formwork of rip-sawn Baltic pine, with boards of varying thickness to give the strong vertical patterning it was desired to achieve. The forms were fixed by screwing from behind, and loose tongues permitted up to twelvefold re-use of some of the boards.

This sample panel was used as a standard by which all other concrete finishes throughout the building were judged; and the successful contractor was obliged by contract to construct a similar sample panel. On these two panels it was possible to solve various problems of finish at an early stage. Thus, calcium deposits were removed by washing down with a weak solution of hydrochloric acid, and a silicone finish was applied to give protection against the weather.

Below: **An external view of the Queen Elizabeth Hall**
(*Photographic Unit, Department of Architecture and Civic Design, Greater London Council*)

The Hayward Art Gallery: South-West aspect from pedestrian walkway

Viewed from the outside, the Arts Centre forms a highly complex whole, with its mass of cantilevered concrete cubes, jutting balconies, terraces and linking walkways piling up like a masterpiece of early Cubist painting.

(*All photographs by the Photographic Unit, Department of Architecture and Civic Design, Greater London Council*)

Seen on the opposite page are (*top*) another view of the Art Gallery, facing west with a corner of the Queen Elizabeth Hall on the extreme left. *Below:* A view of the Lower Gallery, showing the bold internal use of visual concrete.

One of the Sculpture Courts of the Hayward Art Gallery
(*Photograph by Richard Einzig*)

There are three of these open-air balcony terraces, giving a total of 7,000 square feet of display space.

All have wonderful views embracing a great sweep of buildings on the North Bank of the Thames, from the Millbank Tower in the west to St. Paul's and the City of London.

The balconies provide a fine setting for sculptural work of any style or period.

The east wall of the Hayward Art Gallery, showing the fire-escape stairway and the main air-intake duct

(*Photograph by the Photographic Unit, Department of Architecture and Civic Design, Greater London Council*)

Note the pattern of large circular bolt-heads imprinted on the concrete from the carefully-planned formwork. It is a feature constantly repeated, both externally and internally, throughout the Gallery.

The Stand of the Fairydean Football Club, Galashiels, Scotland (1965)

Designer: Peter Womersley, *with* Joe Blackburn
Consulting Engineers: Ove Arup & Partners, Edinburgh
Contractors for concrete structure: Holst & Co. Ltd.

(Photographs by Waverley Studios, Galashiels)

A striking three-dimensional variation on a triangular theme, the stand is an ingenious structure of reinforced concrete based on a module of 5 in.—the width of the dressed Douglas fir boards which comprised the formwork—and on angles of 60° and 30°. It was built for only £25,000—a sum raised by a public lottery.

Rear view of the Galashiels football stand

The stand provides seating for 620 people under cover, with four changing-rooms, showers, clubroom facilities and a kitchen below. The structure, of air-entrained reinforced concrete, consists of four tapering triangular piers, clearly seen in the photograph above, which support the canopy and the back wall of the seating. The canopy cantilevers 15 ft on either side of the end piers, and 26 ft forward over the spectators. Housed in it are four floodlights which, by reason of the slope of the soffit, are automatically angled to light the whole width of the ground in front of the stand.

The concrete back wall of the stand is also triangular in section, and forms one side of the support system for the precast concrete seating units—the other side being provided by the access galleries at either end in front of the stand. A 2 ft 6 in. band of safety glass acts as a shield against wind and rain round the sides and back of the stand.

Noteworthy is the integration into the design of the turnstiles on either side of the stand. Each has its own concrete umbrella on a single support, sheltering the incoming spectators and repeating the triangular theme of the main stand itself.

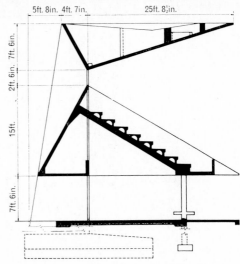

Above: **A cross-section through the structure of the stand**

175

The John Lewis Partnership Warehouse in Stevenage New Town (1963)

Architects: Yorke, Rosenberg and Mardall

Consulting Engineers: Clarke, Nicholls and Marcel (in association with Félix Candela)

Right: **The Loading Dock**
(*Photograph by Henk Snoek*)

The rectangular warehouse covers 145,000 sq ft, and is divided into five bays transversely and fifteen longitudinally. It is entirely roofed with hyperbolic paraboloid umbrellas of pale grey reinforced concrete, each 60 ft by 31 ft and only two inches thick. Co-designer of these umbrellas was the celebrated Mexican architect, Félix Candela. The umbrellas are supported centrally on 13 ft high reinforced-concrete columns of square section, and are tilted to provide north lighting.

Fifteen half-shells, believed to be unique, complete the building along one side. At the loading dock end, three shells were made extra high to accommodate a gantry for high loads.

Externally, the roof is insulated with cork, covered with a light-weight screed and finished with roofing felt. One end of the building is infilled, above a certain level, with greenish glass. Below this glass, and elsewhere, wall infilling is by smooth grey flint-lime bricks.

The entire roof structure was placed with the aid of formwork (four-inch planed boards) sufficient for $5\frac{1}{2}$ shells only. As each row of $5\frac{1}{2}$ shells was completed, the eleven sets of formwork were moved on together to the next position, so giving 15 re-uses by the time the end of the building was reached. The formwork for each half-shell was hinged so that its shallower wing could be dropped to the horizontal when a placing was completed; the entire formwork then needed to be lowered only an inch or two—just enough to release the higher wing from the newly-cast concrete—before it could be moved forward on a bogie-mounted scaffolding frame to its next position.

(*Photograph by S. W. Newbery*)

Concrete furniture in the new Library at Trinity College, Dublin: A view of the second-floor Reading Room

Architects: Ahrends, Burton & Koralek
(*Photograph by John Donat*)

The main upright supports of the study desks were cast *in situ*, with the horizontal wings of precast concrete bolted on to them. The concrete used was cast with white cement and the whitest sand obtainable. The formwork, of British Columbia pine boards, was repeatedly re-used. Note the deliberate pattern of deep bolt-holes left in the concrete.

The Seminary of St. Peter's College, Cardross, near Glasgow (1966)

Architects: Gillespie, Kidd and Coia (*In charge:* I. Metzstein, J. Cowell and A. MacMillan)
Consulting Engineers: W. V. Zinn & Associates
(*Photograph by Crispin Eurich, London*)

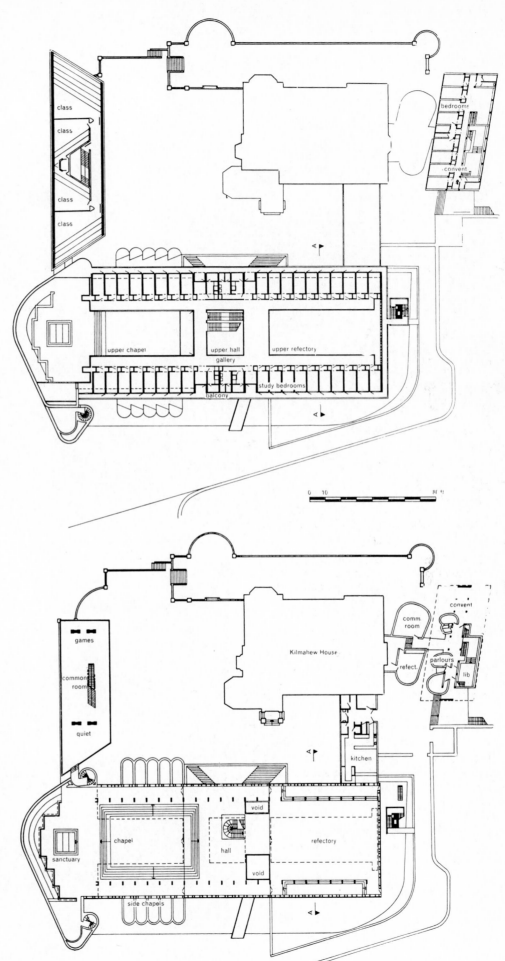

This new seminary for 100 students for the Roman Catholic priesthood is situated some 20 miles north of Glasgow, on the banks of the River Clyde (which can be seen in the background of the photograph opposite).

Built on a steeply-sloping, wooded hillside in the grounds of a sombre Scottish mansion named Kilmahew House, it comprises three main blocks: a long four-storey main block incorporating two chapels, refectory and study-bedrooms; a three-storey library and classroom wing at right angles; and a small two-storey convent wing tucked away behind Kilmahew House in a corner of the site.

The Seminary buildings won the 1966 Architecture Award for Scotland of the Royal Institute of British Architects.

Lighting from the side-chapels of the Cardross Seminary throws up the texture of the external cladding of main block—precast concrete slabs, light brown in colour, with an exposed-aggregate finish of large rounded pebbles.

(*Photograph by Crispin Eurich*)

The photograph to the right (*by S. W. Newbery*) shows the junction between the main block and the classroom wing at right angles to it. The irregularly-spaced mullions and the coarse-pebble exposed-aggregate facing slabs of the chapel contrast strongly with the board-marked visual concrete of the classroom block.

The latter is remarkable for its deep cantilevers and for the boldly-jutting silhouette of its classrooms as they project over steeply falling ground.

The stepped-back structure of the main block derives from a series of combined frames and cross-walls of reinforced concrete, placed *in situ* at 8 ft centres.

A ramp leads from the lower chapel at Cardross, curving up behind the sanctuary with the High Altar seen top right

Other altars at which students exercise themselves in the various services of their rite are seen in the chapel below.

All internal visual concrete was left as it was struck from the forms, for the most part deliberately rough.

The small two-storey convent at Cardross adjoins a corner of Kilmahew House

It is built of double cantilevers in reinforced visual concrete, with precast cladding similar in form and detail to that on the main block.

Interiors throughout the Seminary are in red pine.

The H. J. Heinz Company's new Research and Administration Centre at Hayes Park, Middlesex (1965)

Architects
  In London: Mathews, Ryan and Simpson.
  In New York: Skidmore, Owings and Merrill (Gordon
  Bunshaft in charge of design)
Precast concrete frames: Empire Stone Co., Ltd
(*Photograph by A. Stroud*)

'Up-ended match-boxes' is a term of denigration which has been applied to many large post-war office blocks in Britain; and there are indeed sound reasons, practical and aesthetic, for giving more depth to the façade of a large building. Visual concrete, by virtue of its almost limitless plasticity, lends itself particularly well to two methods of achieving this depth—precast frames and wall panels. Some examples of both are shown on this and the pages following—beginning with this distinguished Research Centre at Hayes Park.

Strong modelling to the tall glass façades of the Centre is given by large cruciform precast concrete frames set 5 ft in front of the glazing. Each frame weighs about three tons. They are tied together by Lee McCall bars, $1\frac{1}{4}$ in. in diameter, housed in ducts running through the vertical members and tensioned as each element was erected.

The joints between the ends of the vertical members occur at eye-level from within the building, so great care was taken to make them precise. Profile tolerances of $\pm \frac{1}{8}$ in. and $\pm \frac{1}{16}$ in. were achieved on the meeting faces between elements. The faces were finished with mild steel plates, 10 in. square, held in exact contact with the adjacent surface by a resin jointing compound.

External members have on all their faces a beautifully even exposed-aggregate finish of a light silvery-grey colour and of very fine texture. White Luxulyan Cornish granite graded to a maximum of $\frac{3}{8}$ in. was used with fines of the same material and white Portland cement. The water/cement ratio was kept within limits of 0·35 to 0·37.

The aggregate was exposed by a combination of washing and etching with acid.

The main entrance to the Administration Building of the new cement works under construction for The Associated Portland Cement Manufacturers, Ltd. at Northfleet, Kent (1970)

Architects: Yorke Rosenberg Mardall, London
Consulting Engineers: Felix J. Samuelly & Partners
(*Photographs by Richard Einzig*)

The site is a reclaimed area of the River Thames on which a new river wall and mooring has been built as part of the new cement works. A system of structural concrete piles supports the building.

Externally, the columns, the spandrel beams and the wall enclosing the upper part of the building are in exposed concrete left from the formwork with $1\frac{1}{2}$ in. vertical ribs projecting at 3 in. centres. These ribs were subsequently knocked off with a club-hammer to expose white calcined flint aggregate—this form of surface treatment being chosen as an economical means of providing a good weathering face to *in situ* concrete and a more consistent finish than could be relied upon from walls left untouched from the moulds.

The first approach to the Architects to discuss feasibility was made in mid-February 1968. The project was completed, furnished and occupied by January 1970.

Wexham Park Hospital, near Slough (1966)

Architects: Powell & Moya (in association with Llewellyn, Davies and Weeks)
Precast concrete: Anglian Building Products Ltd.

(*Photographs by Colin Westwood*)

The $66\frac{1}{2}$-acre site on which this 300-bed hospital is built was once occupied by a Victorian country house, with fine trees and a lake in the gardens. These natural features have been retained in the layout of the hospital, which won an Architecture Award from the Royal Institute of British Architects in 1967.

The planning approach to the hospital was to create a kind of village or small town, capable of growth as the need for more beds grows but with each sector unmistakably forming part of the same organism. The result is a criss-cross of covered, enclosed walkways intersecting below a central 8-storey tower. The different parts of the site have been related and unified by emphasizing the main structural elements, and by adopting a common structural material and finish throughout—concrete, either *in situ* or precast, mostly reinforced but sometimes pre-stressed.

In principle, all superstructure has been expressed, with its exposed members of white concrete made with white cement and calcined flint aggregate, bush-hammered externally to give added texture. The same material and finish was used for the combined flower boxes and seat units of precast concrete which enclose and define the garden courts. The walls below the white superstructure were everywhere rendered a dark coffee-colour, the contrast between the two giving a crisp and lively air to the entire complex.

The dramatic ceiling in faceted, board-marked concrete in the entrance
hall of the Administration Block at Wexham Park

RECEPTION

New Town Hall at Middleton, Manchester (1964)

Architects: Lyons, Israel, Ellis & Partners
(*Photograph by Elsam, Mann & Cooper, Manchester*)

This white concrete building, after several years' exposure to Manchester's notorious climate, is still white—not yellowish or brownish and not, for the most part, streaked. Only on the inward-sloping unwashed facets is it shaded to a silver-grey. Weathering of this type has given (as was intended) depth and character to the building, and is one of its most important design features.

The rough board-markings and rows of indented nail-heads clearly seen in the photograph could suggest that this was a concrete construction cast entirely *in situ*. In fact, only the vertical board-markings on the blank wall surfaces were so cast, all the horizontal panels being part of a carefully worked-out and modelled set of precast cladding panels bolted to the *in situ* frame above and below the windows. Details of this arrangement are shown in the diagram to the right.

Typical panels are 4 ft 9 in. high by 6 ft 5 in. long, and are inclined upwards at the top and bottom to meet the white-painted window frames. At the sides, however, the panels are more nearly in one plane, thereby causing vertical strips of glazing to be set forward from the rest. This has the double advantage of breaking up the external faces of the building and of allowing some windows to be opened inwards without sweeping everything off the broad ledge inside the offices at sill level.

The panels, which are 3 in. thick on the sloping parts and 4½ in. thick in the centre, are reinforced and were cast against rough-textured boards. The nail-head fixings reproduced on the concrete faces in vertical lines sre sunk deep enough to cause marked shadows.

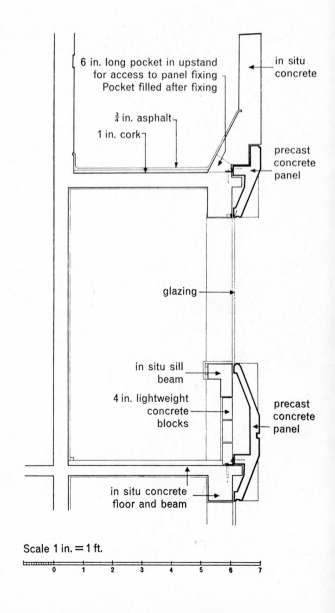

6 in. long pocket in upstand for access to panel fixing
Pocket filled after fixing

in situ concrete

¾ in. asphalt
1 in. cork

precast concrete panel

glazing

in situ sill beam

4 in. lightweight concrete blocks

precast concrete panel

in situ concrete floor and beam

Scale 1 in. = 1 ft.

0 1 2 3 4 5 6 7

The United States Embassy, Dublin (1965)

Architects: John M. Johansen (New York) in association with
Michael Scott and Partners (Dublin)
Precast concrete frames: Schokbeton Products Corporation,
Kampen, Holland

(*Photograph by S. W. Newbery*)

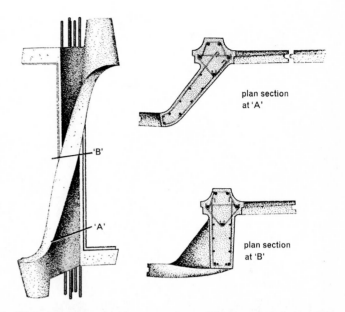

plan section at 'A'

plan section at 'B'

The sinuous precast concrete frames which cover the curved façade are repeated round a galleried rotunda in the centre of the building. In his drawings for the twisting vertical units (*of which a typical one is seen on the right*), the architect provided nine plan sections cut at equal intervals down the height of the unit. Their manufacture still presented a difficult task, however, for an error of only $\frac{1}{8}$ in. in each external unit would have meant an overall error round the circle of 3 ft 6 in.

The mould was constructed of fibre-glass laminations stiffened with plywood supports. White cement was used in the making of the 'no-slump' concrete, with only enough water to enable the chemical reaction to be completed. The concrete was then vibrated ('shocked') on a table moving $\frac{1}{4}$ in. up and down 250 times a minute. Later, the units were bush-hammered, and shipped from Rotterdam to Dublin.

A Block of study-bedrooms at St. Anne's College, Oxford (1964)

Architects to the College: Howell, Killick, Partridge and Amis
Precast concrete units: Trent Concrete Ltd.

(*Photograph by S. W. Newbery*)

Built and furnished with a donation of £100,000 from the Isaac Wolfson Foundation, this is the first block in what is planned to be the rebuilding of the greater part of the College. Lozenge-shaped in plan, it has highly modelled façades which provide, in the words of Nicolette Franck (M.A. Oxon.), "a joyous affirmation of how pleasing concrete can be as a surface material, either board-marked direct from the forms or in the form of exposed-aggregate precast units. . . .

"The side walls of each room are splayed out by 4° so that a range of rooms together forms a curve, and builds up into a block with highly modelled convex sides and angled end walls. . . . This allows space at the centre of the building for a generous helical staircase . . . formed of precast concrete units each comprising one step and a central core ring, placed on top of one another so that the rings made permanent formwork for an *in situ* reinforced concrete column.

"The ground floor takes the form of an open colonnade on the inward-looking side, and of a solid, cantilevered plinth over a dished lawn on the (other) side. Eventually a lake will be formed here, and the ground-floor rooms will cantilever out over it."

All the *in situ* structural concrete—with the exception of the floors—is board-marked straight from the forms, which were wire-brushed to raise the grain. As the concrete had to blend with adjoining stone buildings, local yellow pit sand was introduced into the mix to give a creamy colour when combined with white cement.

The balcony units, which weigh $2\frac{1}{2}$ tons each and were cast in one piece, were fixed in position with *in situ* joints at floor level. All the cladding units were cast solid, face down in timber moulds, using the facing mix throughout. Where the aggregate was exposed, the mould surfaces were painted with a retarder and the concrete surface was brushed after the mould was stripped. The mix for the cladding combined Derbyshire spar, De Lank Cornish granite, Leemoor sand and white cement.

Centre Point, London (1965)

Architects: R. Seifert & Partners
Engineers: C. J. Pell & Partners
Contractors: George Wimpey & Co. Ltd.
Precast concrete units: Portcrete Ltd.

This office tower block rises dramatically 370 ft over the junction of Oxford Street and Tottenham Court Road. It has two particular assets: its graceful profile when viewed from a distance, and the strongly-modelled precast white concrete units which give a deep reveal to the office windows on all its façades.

This latter feature is of special value on tall blocks, preventing vertigo and giving some shade to the windows.

189

The Cumberbatch Extension to Trinity College, Oxford (1965)

Architects: Robert Maguire and Keith Murray
Associate: Gordon Russell
Structural Engineer: W. H. Aubrey & Partner
Contractors: Bovis Ltd.
Modular Concrete Blocks: Forticrete Ltd.

(*Photograph by Keith Murray*)

The extensions to the Jacobean buildings of Trinity are known as Cumberbatch North (*see above*) and South Buildings. They face one another across a large open quadrangle paved with circular polished-concrete flagstones set in patterns of pebbles in mortar. The North Building has a basement lecture-theatre seating 50, a formal reception room at ground floor, and three storeys above containing 15 two-room sets for graduate or undergraduate occupation.

Junction of new modular concrete blockwork and Jacobean stone

Both buildings of the Cumberbatch Extension at Trinity are constructed of board-marked reinforced concrete slabs supported on load-bearing concrete block walls and (in places) concrete columns. Blue engineering brickwork at ground level: otherwise (and perhaps the main feature of technical interest in the Buildings) external walls are built of modular concrete blocks, each an exact cube of eight inches joined to its neighbours above, below and on either side by raked joints.

The photograph to the right (*copyright Architectural Press*) shows how well this concrete blockwork marries with the Jacobean stonework, even before cleaning. The warm grey of the blocks—varying slightly in colour one from another, with consequent enlivening effect on the appearance of the walls (*see last page*)—harmonizes even better with the buff tones of the old Clipsham stone of Trinity Library now that the latter has been washed.

The Cumberbatch South Building at Trinity College, Oxford

(*Photograph by Keith Murray*)

The modular concrete blocks of 8 in. cube have the architectural virtue of relatively small scale. Blocks of this size look well in a domestic or (as here) semi-domestic context, weathering evenly and giving walls built of them a pleasant texture overall. As building units, the relatively small blocks are convenient to handle, and allow much flexibility in design.

The Cumberbatch South Building is L-shaped, partially enclosing a smaller brick-paved quad. It contains further sets of undergraduate rooms in three storeys, with a large trunk store above.

St. Michael's College Chapel, Kirkby Lonsdale, Westmorland (1966)

Architects: Building Design Partnership
(*Photograph by Elsam, Mann & Cooper, Manchester*)

An original method of converting standard-size concrete paving slabs into building blocks of vigour and distinction was developed by Forticrete Ltd. for both internal and external use in the Chapel, a general interior view of which is seen above. The Chapel stands in the grounds of Underley Hall, a Jacobean-style mansion used as a Junior Seminary for the Roman Catholic priesthood. A material was sought which would blend in both colour and texture with the weathered masonry of the Hall— but at a price well below that of quarried stone.

After much experiment, a concrete mix was achieved which, consisting of 1 part sand, 4 parts Dyserth lime dust and 6 parts cream-coloured Breedon limestone aggregate of $\frac{3}{8}$ in. screen, gave a good colour match; but the problem of texture remained. Two basic requirements were that the blocks must have three split faces, to form projecting nibs; and that the height of the courses had to be limited to a maximum of 3 in. to achieve the effect which the designers had in mind. Standard blocks, untreated, would have provided only two split faces

and a minimum course height of 4 in.

So blocks were hydraulically pressed, before they had set, to a uniform thickness of $2\frac{5}{8}$ in., and then split longitudinally and transversely into building units of sizes suitable for the various structural functions of the walls. The splitting process produced blocks varying in size from 16 in. × 8 in. to 8 in. × 4 in.—but all of the standard $2\frac{5}{8}$ in. depth. Cavity walls were formed with an outer leaf of the 4 in. blocks, a two-inch cavity, and an inner leaf of 8 in. blocks.

Since the ends of every block were left naturally rugged by the splitting process, all vertical joints were made flush. Horizontal joints between courses were raked out and then rounded with a finishing tool, so forming longitudinal shadowed recesses which act as a foil to the vertical emphasis of the shafts that form the principal features of the building.

The unusual dimensions of the building units to be used led the architects to prepare, at the design stage, bonding sketches which were sent out with the invitations to tender. Full working drawings for the bonding pattern of every wall were later supplied to the successful contractor.

The concrete block-making factory operated by Forticrete Ltd. at Aintree, Lancashire

In the factory, all the batching and mixing processes (especially measurement of the exact quantity of water needed to give the desired workability) are electronically controlled to critical limits.

Crushed granite and limestone aggregates, sand and cement are unloaded in the factory yard; and the sand and aggregates are raised by conveyor-belt to four hoppers in the factory roof. The cement is blown to a fifth hopper, while a smaller dispenser holds the metal oxide powders used to impart a wide range of colours to the mix, and the aluminium stearate used to waterproof the final product. Two mixers below these hoppers discharge into a skip, which is then raised from basement to roof to pour the mix into the feed hopper of the blockmaking machine. From there it is fed into a large steel mould precision-tooled for every type of block or brick required.

A "head" comes down on top of the concrete in the mould, and the mix is vibrated down to the required height before being pushed out on to a steel pallet. A fork-lift truck takes loaded pallets into one of six low-pressure steam-curing kilns, where live steam at 150°F circulates evenly through the blocks for 17 hours.

The cured blocks are unloaded automatically on to a roller conveyor, for stacking by a second fork-lift truck in the yard. The steel pallet on which they were moulded and cured is fed back into a magazine ready to re-start the manufacturing cycle.

A single operator supervises the automatic moulding machine in the factory

Moulded blocks pass to the left on their way to the kilns; steam-cured blocks roll down the conveyor at the bottom. Six men on an eight-hour shift can mould and handle 20,000 blocks. A special machine splits some of these into smaller sizes and different shapes, giving them a natural ruggedness of profile which makes them of great interest to architects.

A staircase in St. Michael's College Chapel, Kirkby Lonsdale

(*Photograph by Elsam, Mann & Cooper, Manchester*)

Top-lighting throws into high relief the textured concrete blockwork of which the Chapel is built.

Radio Mast for the Headquarters of the Durham County Police, just outside the City of Durham (1969)

Consulting Engineers: Ove Arup & Partners (*with* J. L. Parnaby, ARIBA, Architect to the Durham County Council)

Contractor: Bierrum & Partners Ltd.

The consulting engineers developed an attractive three-dimensional design in which a slender mast is superimposed on a tripod. Their purpose was to avoid a guyed structure yet retain sufficient stiffness to limit tilt and induced oscillation under wind loading. At a wind speed of 50 m.p.h., maximum angular rotation of the mast in the vertical plane is less than $\frac{1}{2}°$.

(Photographs by Turners (Photography) Ltd., Newcastle-on-Tyne)

The radio mast under construction

The structure was precast on site in five elements: the 119 ft mast itself, the key, and three 60 ft units. In this way the need for full-height scaffolding and formwork was avoided—as was the need to protect the temporary construction against wind.

The concrete was specified to have a 28-day strength of 4500 lb per square inch, using white Portland cement. The whole of the structure was left smooth, with a fair-face finish. It cost in the region of £10,000.

The design won a Special Mention in the 1969 Award of the Concrete Society, by courtesy of whom the photographs are shown.

Bridges over the Yorkshire Motorway (1968)

Fifteen of the bridges over the Motorway, submitted as a
group entry, received special mention from the judges
for the Concrete Society's 1969 Award. Design was by the
Highways & Bridges Department of the West Riding of
Yorkshire County Council, acting as Agents for the
Ministry of Transport.

(Photograph by Elsam, Mann & Cooper)

(Photograph by The Daily Telegraph)

Above: **The Needle Eye Accommodation Bridge**
A three-pin arch with side cantilevers, 16 ft 4 in. wide and
290 ft between springings. Height above Motorway
at crown: 53 ft.

Left: **The Droppingwell Footbridge**
A steady gradient of 1 in 10 is maintained throughout this
elegant bridge, which springs across the Motorway in a
170 ft span and coils down gently on the other side.

Below: **The Smithy Wood Footbridge**
Believed to be first concrete bridge to employ the Wichert
Truss principle—the pinned end members over the internal
piers forming open rhomboids, so permitting differential
settlement of any support and relieving mid-span moments.
The area is subject to coal-mining subsidence.

(Photograph by Elsam, Mann & Cooper)

Taf Fechan Bridge on the Heads of the Valleys road, Glamorgan (1964) built for The Welsh Office, Roads Division

Consulting Architects: Alex Gordon & Partners
Consulting Engineers: Rendel, Palmer and Tritton
(*Photograph by William Tribe Ltd., Raglan*)

This 392 ft bridge throws twin parabolic arch ribs of reinforced concrete 228 ft across the gorge of the Taf Fechan stream, yet these ribs are only 3 ft 7½ in. thick at the springings and 2 ft 6 in. thick at the crown. Each rib is 13 ft wide, and is straight—even though the deck of the bridge is curved on plan. The required curvature is obtained by varying the width of the cantilevered portion of the deck.

The provision of formwork and scaffolding for big arch bridges presents a considerable problem, and this was a major factor in the decision to build the bridge by an unusual system combining cantilever construction with temporary prestressing. The two arch ribs were constructed without scaffolding by cantilevering out from the abutments towards the centre. The ribs were supported during construction by cables with ½ in. diameter strands, anchored to temporary concrete towers erected on the bridge deck over the arch springings. The forces in the towers were balanced by cable stays anchored in the arch abutments. Temporary stressing and de-stressing of the cables was carried out as the arch progressed in order to obtain the required changes in suspension forces.

The concrete for the arch rib was placed on travelling formwork, which was supported by a section of the rib already constructed. Each half of the arch was completed in five major placings. Short gaps were left at the crown just long enough to allow for the reinforcement to be spliced, and were then concreted.

'The Mermaids': Sculpture in concrete in the grounds of the Cement and Concrete Association Research Station and Training Centre at Wexham Springs, Slough

Designer: Sheila Gabbe

(*Photograph by Sheila Harrison*)

These two seated mermaids, whispering to one another forever of the sea, were the product of a complicated piece-mould in plaster taken in 78 pieces off an original in bronze. Sculptors Ltd., who cast it for the artist, say of the design that, though not an ideal one for casting in concrete, it proves that even the most complicated composition of linked but varying volumes can be successfully cast in this wonderfully flexible material.

The mix used was 1:3 rapid-hardening cement and washed Thames grey sand, maintained at a hand-workable consistency throughout. The separator between mould and concrete was a mixture of shellac and tallow.

The surface of the sculpture was finished with wax and carbon black to give it an even dark-grey colouring overall.

Concrete statuary at Wexham Springs

Also in the grounds at Wexham Springs stands Lester Dealtry's impressive 'King'. This large piece, weighing 30 cwt, was cast entirely in the artist's studio. Modelled and moulded in the vertical, the plaster mould was then filled horizontally with a quarter-inch-thick surface mix of 1 : 3 white cement and washed Thames grey sand. The concrete was compacted by hand, and 'The King' was reinforced throughout with half-inch round bars. He needed to be lowered by crane on to steel uprights set in concrete when he arrived at Wexham.

The separator between mould and concrete was a mixture of shellac and tallow.

(*Photographs by Sheila Harrison*)

In the yew arbour in the gardens at Wexham Springs stands Edward Folkard's life-size concrete 'Nude'

Head turned to look past her right shoulder and clasped hands resting on her right hip, she was cast in a plaster mould and reinforced, then made with rapid-hardening cement of a dark colour characteristic of this artist's work.

Chimney Breast in the Officers' Mess, St. Patrick's Barracks, Ballymena, Ulster

Design: Desmond Kinney and Ralph Dobson
Architects: James Munce Partnership

A young design group in Belfast has been sketching murals in concrete which forward-looking architects in Northern Ireland have been commissioning for hospitals, hotels, office blocks, public-houses, shops, churches—and a brewery.

Weapons and armour form the appropriate basis of the design of the chimney-breast in the Officers' Mess above. The 3 in. thick concrete panels were cast against polystyrene moulds, and bolted to the wall. The pewter colour of the finished work was obtained by applying silver powder mixed with white spirit. When dry, the wall was given a final polish with black-tinted beeswax.

The 9 in. thick concrete mantelpiece was cast separately and fitted to the assembled panels.

Left: **A wall in the entrance hall of the Pig Marketing Board, Belfast, by the same archtects and design group**

(*Photographs by S. W. Newbery*)

(Photograph by Photo Finishers (Sheffield) Ltd.)

Mural in the entrance to the St. John Fisher Secondary School, Woodhouse, Sheffield

Designer: Philippa Threlfall
Architects: Hadfield, Cawkwell, Davidson
            and Partners

An example of a new form of mural in visual concrete, subtle and restrained in design and simple to execute. Precast concrete slabs, scored on their upper surface, are made in sizes which make them fairly light to handle yet thick enough to support their own weight (2 ft square by 2 in. thick is typical).

Forms are placed around each slab, to receive a one-inch rendering on the scored surface. Into this rendering are then embedded any decorative pieces of material that come to hand—either natural such as stones, pebbles or fragments of rock, or manufactured such as terracotta or bits of ceramic. The finished slabs are secured to the wall by metal ties.

Murals of great richness have been achieved by this young artist, varied in texture and often subtle in colour. Though the general design is always thought out beforehand, the details of the mural develop as the work proceeds—thus giving a freedom of design which could never come from the alternative technique of arranging the decorative materials face downwards in the bottom of a mould and casting the slabs on top.

Joints between every slab in Miss Threlfall's murals are clearly expressed, and are often used as a unifying factor in the design or as a convenient means of dividing up and organizing space.

A wall at the North London Collegiate School, designed and executed by students working under the supervision of Miss Threlfall (1965)

(*Photographs by Sheila Harrison*)

The mural covers one side of an outdoor wall separating two car parks. The theme is a song of praise introducing a wide range of birds, fish, animals and people, so giving maximum scope for invention. Its details were allowed to evolve in accordance with the pupils' own ability and imagination.

Stones were collected from all over the United Kingdom. Some came from the Continent of Europe and from America, and one arrived from Mount Zion itself. Pupils applied the rendering, made and glazed the ceramic pieces, set them and the stones in the rendering, and pointed up the work.

Additives were used to darken the concrete and to bring out the richness of the stones.

Right: **Detail of the Mural**

Mural displayed in the grounds of the Research Station of the Cement and Concrete Association at Wexham Springs near Slough

Designer: William Mitchell

(*Photograph by A'Court Photographers Ltd., Staines*)

The mural is built up of concrete units, each precast against clay moulds and coated with polyester resin either clear or pigmented according to the requirements of the overall design.

Below:

Concrete murals in the Banking Hall at Martin's Bank, Watford

Designer: Eric Peskett
Architects: Bryan and Norman Westwood & Partners

(*Photograph by Colin Westwood, Weybridge*)

Square plaques cover a series of concrete louvres running from floor to ceiling to form a complete wall in the banking hall. Each plaque is decorated with a relief motif cast with high-alumina cement and light-weight aggregates against rubber moulds lying on polished Formica.